Synthesis Lectures on Engineering, Science, and Technology

The focus of this series is general topics, and applications about, and for, engineers and scientists on a wide array of applications, methods and advances. Most titles cover subjects such as professional development, education, and study skills, as well as basic introductory undergraduate material and other topics appropriate for a broader and less technical audience.

Deandra Cassone

Operational Leadership Using Quantitative Decision Making

Springer

Deandra Cassone
Kansas State University
Manhattan, KS, USA

ISSN 2690-0300 ISSN 2690-0327 (electronic)
Synthesis Lectures on Engineering, Science, and Technology
ISBN 978-3-031-67564-5 ISBN 978-3-031-67565-2 (eBook)
https://doi.org/10.1007/978-3-031-67565-2

This Springer imprint is published by the registered company Springer Nature Switzerland AG
The registered company address is: Gewerbestrasse 11, 6330 Cham, Switzerland

Preface

Throughout my career, I have had the privilege of being exposed to superior business leaders, academics and mentors and have welcomed every opportunity to learn from them. This book captures a part of the knowledge that I've gained from these individuals and applied in my own career leading to various accomplishments.

Dr. Ching-Lai Hwang was a professor in the Department of Industrial Engineering at Kansas State University. Dr. Hwang had a great mathematical ability, studied methods, and published numerous books and papers in the field of multiple objective decision making, multiple attribute decision making and group decision making as well as developing his own well-recognized multiple attribute method called the Technique for Order Preference by Similarity to the Ideal Solution (TOPSIS). He and his graduate students solved problems using the methods that were presented in each of his books.

Dr. Frank Tillman was head of the Industrial Engineering department at Kansas State University. Dr. Tillman had a strong mathematical, statistical, industrial engineering, finance and business background and was able to model a broad variety of problems by assessing aspects of problems, applying quantitative methods to the problem components, and aggregating the components into a model that represents a problem solution from a holistic perspective. These models provided robust solutions to real-world problems.

Dr. Tillman and Dr. Hwang were asked numerous times to solve problems for former students in the military and in industry and received a contract and investment that enabled them to start two consulting companies, one for private-sector projects and the other for government and military projects. After receiving several contracts and taking a year's sabbatical from the university, Dr. Tillman began running the consulting companies full-time. Dr. Hwang stayed in part at the university and in part as a principal of the consulting companies. The consulting companies used the techniques and approaches described in this book to solve industry, government, and military problems successfully.

After working in the mentioned consulting companies for nearly 20 years, I went back to school to pursue my Ph.D. It was an excellent opportunity to gain additional insights into data, methods, and the theory behind the applications we developed. I was 42 years old when I pursued my Ph.D. in Industrial Engineering. One of my committee members, a well-known and well-published professor, referred to me many times as "applied" and an "excellent" student. I view myself as an "understudy" of the "masters" that I mentioned above. Having been exposed to and applied the approaches described in this book, I have attempted to leverage and expand upon the thought processes we used to model and solve complex problems.

After completing my Ph.D., I took a position at a high-tech firm. There I was able to apply this modeling approach to develop twelve awarded business process patents and receive an innovation leadership award at the company that is given to only one employee each year. I attribute this to studying the approaches that we used in our consulting companies and learning the application of tools and techniques as done by Dr. Tillman and Dr. Hwang. Being a consultant earlier in my career led me to ask why certain methods were used and combined to solve problems. The fundamental answer was that the methods selected addressed a specific component of the problem but many times, multiple methods were typically needed to solve a problem. There was really no one-size-fits-all approach to modeling real-world problems. I saw that the quantification of the data, application of methods and integration of decision maker's perspective was important in the problem-solving process. The unique use of methods in various situations was also interesting and, in many ways, provided a freedom to appropriately use methods where they may not have been used previously.

Dr. Tillman and I wrote a book, *A Professional's Guide to Decision Science and Problem Solving* and published it along with four of our consulting studies in 2011 and 2012. This book contains some of the same components that were presented in that book, but focuses on, in many ways, what was inherently known by Dr. Tillman and Dr. Hwang but was not drawn out as specifically in the other books. What was not as apparent in the other book was the essential knowledge of a broad group of quantitative methods and how those quantitative methods can be integrated into quantitative decision-making processes. I saw something great happen with modeling real-world problems with robust analytical and quantitative techniques and am attempting to communicate that thought process beyond our original books. Dr. Tillman and I were discussing the potential contents of this book prior to his passing but due to health issues, were not able to get started. I promised him that I'd write this book.

There is nothing "gimmicky" about this book. To holistically and robustly model problems, you need to have a good understanding of your environment, your data, your decision makers and methods that could and should be applied to your operating environment. Quantitative methods are only presented at a high level in this book. The great thing is that there are many experts in various fields that you can engage to help develop the specifics supporting quantitative decision making. Consulting firms, internal experts, university professors, graduate students and beyond have this additional exposure to the application of methods and can be engaged to assist in the development process. I hope that you find this book helpful in your quantitative decision-making processes.

Manhattan, USA Deandra Cassone

Contents

About the Author

Deandra Cassone PhD, PMP, recently served as the Director of Research Activities, Biomedical Informatics at the University of Missouri and the Administrative Director of the PCORnet Clinical Research Network for the Greater Plains Collaborative. She received a B.S., M.S. and Ph.D. in Industrial Engineering from Kansas State University. Her twenty years of consulting experience involved working with Fortune 500 companies, government agencies, and the military applying decision science and analytical methods to solve complex problems. Her ten plus years of experience working for Fortune 100 companies includes serving as a subject matter expert and in management positions in domain areas such as technology development, logistics operations, operational excellence, strategic project management, and manufacturing operations. She has been awarded twelve business process patents and a corporate-wide innovation leadership award. Dr. Cassone has developed and taught eleven different graduate engineering courses over the last fifteen years as an adjunct professor for Missouri University of Science and Technology and a full-time professor at Kansas State University. Currently, Dr. Cassone is an Adjunct Senior Professor of Practice at Kansas State University and is the president of consulting and investment businesses. She has also published several books in the field of applied decision science.

Key Components of Quantitative Decision Making

Abstract

This chapter introduces an executable approach to developing quantitative decision-making solutions. Leaders focused on robust data analysis, methods application and engagement from decision makers enhance their ability to solve problems and deliver results. The general approach presented includes three components. Exploration discusses the need to have a holistic view of the environment and measure success in accomplishing the objectives of the organization. Quantitative Decision Modeling discusses data mining and statistical analysis, identifying operational improvement methods, applying decision methodologies and selecting the best method supported by the data. Implementation discusses quantitative decision modeling and automating the decision process using robust project management techniques along with monitoring the success of accomplishing the objectives. This chapter also discusses the role of the operational leader in minimizing costs, maximizing service delivery, and delivering quality products. Tools, methods, systems, and processes are required to keep a business running on a day-to-day basis. Operational leaders must also interface with senior executives to meet the organizational objectives. Operational decisions impact organizational objectives. This chapter introduces the operational leader to quantitative decision-making processes and the use of robust analytics and tools to assist the operational leader in performing their job and provide quantitative insights into their organization's performance.

D. Cassone, *Operational Leadership Using Quantitative Decision Making*,
Synthesis Lectures on Engineering, Science, and Technology,
https://doi.org/10.1007/978-3-031-67565-2_1

Introduction

Problems present themselves in an organization on an ongoing basis. You may have issues with customer service, product development, financial tightening due to limited resources, quality issues and many, many more. When problems arise, many organizations will approach solutions from a less than comprehensive perspective. A well-meaning functional division may develop solution approaches that solve their immediate issues, however, impact other functional divisions negatively. Some problems are well solved within your function, however, problems with cross-functional impacts should be explored, assessed, and solved from a holistic organizational perspective. Organizational leaders should assess and solve issues with other functional organizations in mind to minimize negative impacts to other functional organizations. Leaders focused on robust data analysis, methods application and engagement from decision makers enhance their ability to solve problems and deliver results.

This book provides an approach to holistic problem solving and focuses on the identification and use of quantitative methods to generate solutions and improve operations. Quantitative methods provide leadership with defensible tools for decision making.

To strategically solve problems and move an organization in its desired direction, the problem must be explored from an organizational perspective, key leaders must be engaged, sound methods must be used, and implementation methods and metrics must be tracked. There is typically no quick fix in solving complex and critical organizational problems and leaders should be prepared to spend the time and energy to find the best organizational solutions.

The book describes a framework that starts with developing a holistic view of the organization and then defining the organization's goals, objectives, decision criteria and metrics to result in meaningful measures for organizational decisions. Organizations are looking to maximize the impact of their resource dollars and this approach provides an executable process that aligns corporate decision making with corporate objectives. It is important to take time to understand your organization and what is strategically important to your leaders.

Decision-making tools, analytical methods and optimization techniques enhance the objectivity and robustness of your decisions. This framework supports the use of these tools and enables you to make decisions that drive bottom-line performance. This framework can be used by cross-functional executives that impact portfolio management, process improvement implementation, project selection decisions, product development decisions, resource allocation decisions and other critical decisions.

Most decisions faced by management are based on several conflicting goals and objectives and a set of conditions which must be satisfied. This problem-solving approach incorporates experience and expert judgment along with statistical data and quantitative methods to solve multiple criteria problems and perform complex trade-off analysis, and it greatly facilitates individual and group decisions, as well as consensus building at all stages of the decision process. Ultimately, it is key organizational leaders that make these decisions, so approaching organizational leadership with a quantitative decision-making perspective engages leaders in the decision process from problem definition to solution generation. This approach is applicable to most organizations and companies and has evolved from solving a broad group of problems for commercial, government and military organizations. Figure 1 shows a high-level view of this process.

The book is divided into three sections. The first section addresses the high-level exploration of the problem and developing quantifiable measures to ensure decisions are meeting corporate objectives. The second section focuses on the analytical and decision-making processes that can be used to make decisions and solve problems using quantitative methods. The third section provides insights into process implementation from a solution identification, project management and project tracking perspective. The book sections and topics as shown in Fig. 1 are outlined below.

Part 1—Exploration.

- Holistic View to Identify the Right Problem
- Measuring Success in Accomplishing the Objectives

Part 2—Quantitative Decision Modeling.

- Data Mining and Statistical Analysis
- Identifying Operational Improvement Methods
- Applying Decision Methodologies
- Selecting the Best Method Supported by the Data

Part 3—Implementation.

- Modeling and Automating the Decision Process
- Monitoring Success in Accomplishing Objectives

Decision makers must be involved in all phases of structured problem-solving and model development. First, the key decision makers involved in the process must be identified. Then, a group consensus of the objectives that are to be accomplished for organizational success should be agreed upon by these decision makers. This involves bringing together key individuals from various levels of the organization and the functional areas and developing a group consensus on the goals and objectives of the organization.

As part of developing an understanding of the problem and the objectives to be accomplished, key individuals should also help define the decisions to be made, the business environment in which they are operating, the constraints in which the decision process must function and the key decision or variables that are the output of the decisions.

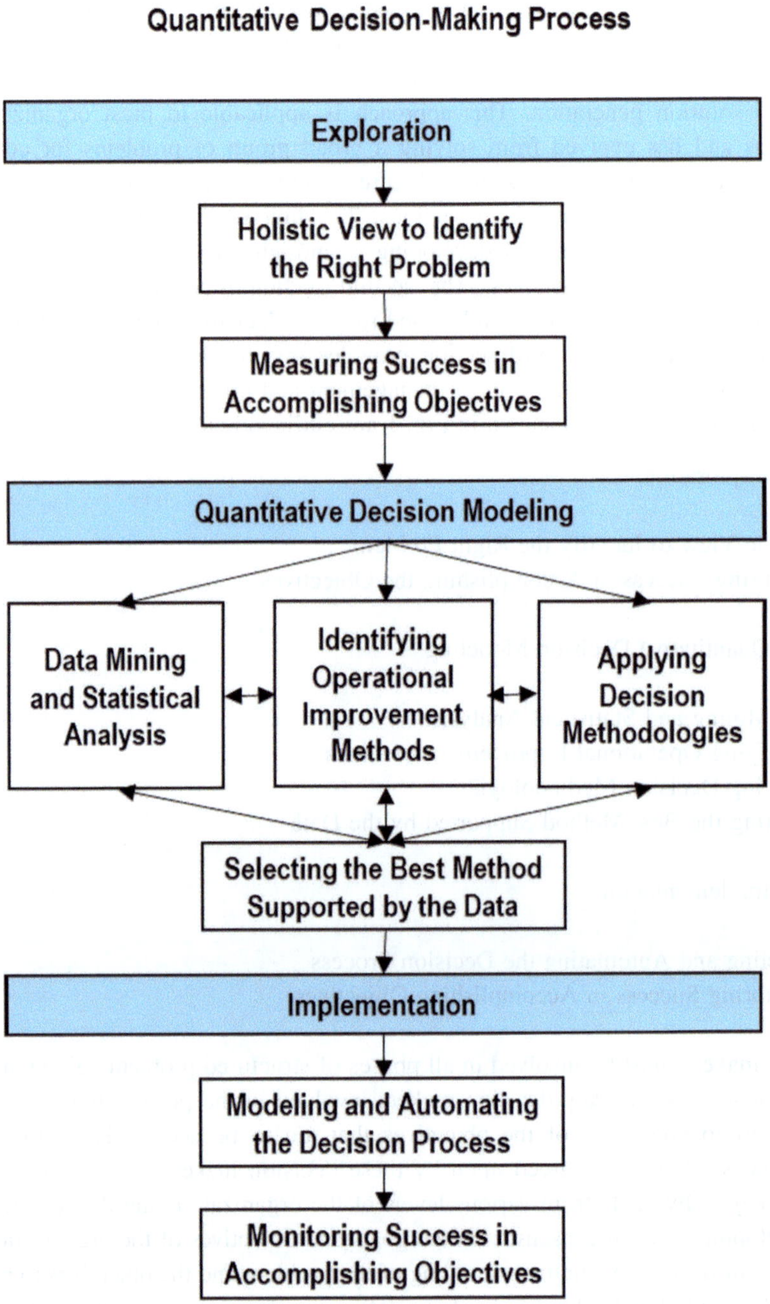

Fig. 1 Quantitative decision modeling

The decision makers then must determine how these objectives can actually be measured within the organization. This might include determining what measures would be use when you want to "reduce cost." They may include manufacturing costs, distribution costs or other important costs. The establishment of metrics to measure the objectives is absolutely necessary because this provides the means to quantify the degree of success in accomplishing the objectives.

With the basic problem defined and structured, data is gathered and analyzed statistically to understand the nature of the information available and the operating performance of the organization. This data analysis is useful in identifying conditions or trends within the organization and is evaluated for its potential usefulness in developing the decision process.

Methodologies are then reviewed to determine what methods best fit the problem at hand with the available data. The methodologies are then selected which best fits the data available, the business constraints, the decision to be made and the objectives and goals to be accomplished. Along with an understanding of the organizational environment, the information that the data shows, and robust quantitative methods, you must include cross-functional decision makers in the decision process. Quantitative decision-making tools can be employed to integrate decision makers in the process. This ensures that your problem solution will be meaningful and used.

The best method for modeling the decision process is then selected. The views of the decision makers are constantly changing. Therefore, it is necessary for the model to easily change over time. This includes building into the model the ability to adapt the decision process to represent the changing strategic direction of the organization. This may seem like a tall order for a decision model, but this book presents an integrated approach that can be used to incorporate these key components in the problem solving and decision-making process. It can be a balancing act to provide leadership with an analytical framework in a manner that can be readily used and avoids unnecessary complexity.

For the maximum usefulness of a decision process, it generally should be automated so that changes can be made quickly and easily. This enables decision makers with the ability to run analyses and update model parameters easily and rerun the model quickly for new answers. Once automated, the decision makers can run various test scenarios, "what-if" analysis and sensitivity analysis to generate decisions that best fit the objectives of the constantly changing business environment. You can also use the automated process to measure and monitor results.

Key in modeling the decision process is ensuring that the approach taken is one that provides a solution that is better than what is currently being done. The best methods to use in modeling are the methods the decision maker will actually use and that he/she can understand. The decision makers must have faith in the decision process because these solutions affect the future of their companies and their own careers. They must understand the method before they use it, so it needs to be simple as possible and straightforward [1, 2].

Operational Leadership and Quantitative Decision Making

Indeed.com defines operational leadership as "the management, development and implementation of business operations. It involves overseeing employees, production and financial costs. This leadership style typically requires problem-solving, critical thinking, analytical skills and communication skills" [3]. Operational leaders focus on managing and improving core business functions such as safety, cost, quality, and service delivery. These key business functions make an organization an ongoing business concern and lead to its success. As such, operational leaders are typically involved in the day-to-day functioning of a business and serve as an interface between the "factory floor" and senior executives.

Tools, methods, systems, and processes are required to keep a business running. Operational leaders must interface with senior executives to meet the organizational objectives. Operational decisions impact organizational objectives. Many times, operational leaders are focused more on their operational responsibilities, especially since their job performance is driven by operational success, however, they may not explicitly associate their operations with strategic objectives. This book assists operational leaders in bridging that gap and approaching operations from a strategic perspective.

The book focuses on quantitative decision making to assist the operational leader in performing their job and provide quantitative insights into organizational performance. There are not only quantitative methods that are used to run operations, but also quantitative methods that are available to make decisions. Data and operational methods support day-to-day operations. Decision-making methods ensure that the decisions that are made support the strategic direction of the company. These methods in combination provide robust operational and decision-making processes to enhance overall organizational operations. Quantitative methods that enhance both the objectivity of assessing operations and the objectivity of decision-making provide insights and defensibility for the decisions made in an organization.

References

1. Cassone, Deandra T., Tillman, Frank A., *A Professional's Guide to Problem Solving with Decision Science: An Integrated Approach for Assessing Issues, Finding Solutions, and Reaching Corporate Objectives*, Second Edition Overland Park, KS, Pioneering Partnerships, LLC, January 2018.
2. Tillman, Frank A., Cassone, Deandra T., *A Professional's Guide to Decision Science and Problem Solving: An Integrated Approach for Assessing Issues, Finding Solutions, and Reaching Corporate Objectives*, Upper Saddle River, NJ, Financial Times Press/Pearson Education, Web: December 1, 2012, Hardcover: February 2012.

3. Indeed Editorial Team, What Is Operational Leadership? (Importance, Duties and Traits), Updated June 24, 2022, Link: What Is Operational Leadership? (Importance, Duties and Traits). Indeed.com.

Exploration

Part I discusses the exploration phase of the process. Concepts discussed include gaining an understanding of your operating environment and defining organizational goals, objectives, decision criteria and metrics to evaluate your decisions. The concepts discussed in these chapters assist you in ensuring that you've captured key considerations in your environment and that your decisions align with your organization's goals and objectives. Figure 1 shows this as the Exploration phase of the process. This part sets the framework for quantitative decision making and engages corporate leadership to ensure that the activities that your organization pursues align with the strategic direction of the organization. Methods and approaches discussed in this section have been used in quantitative decision making consulting engagements and have proven beneficial to engage key leaders in reaching consensus and implementing a quantitative decision making process.

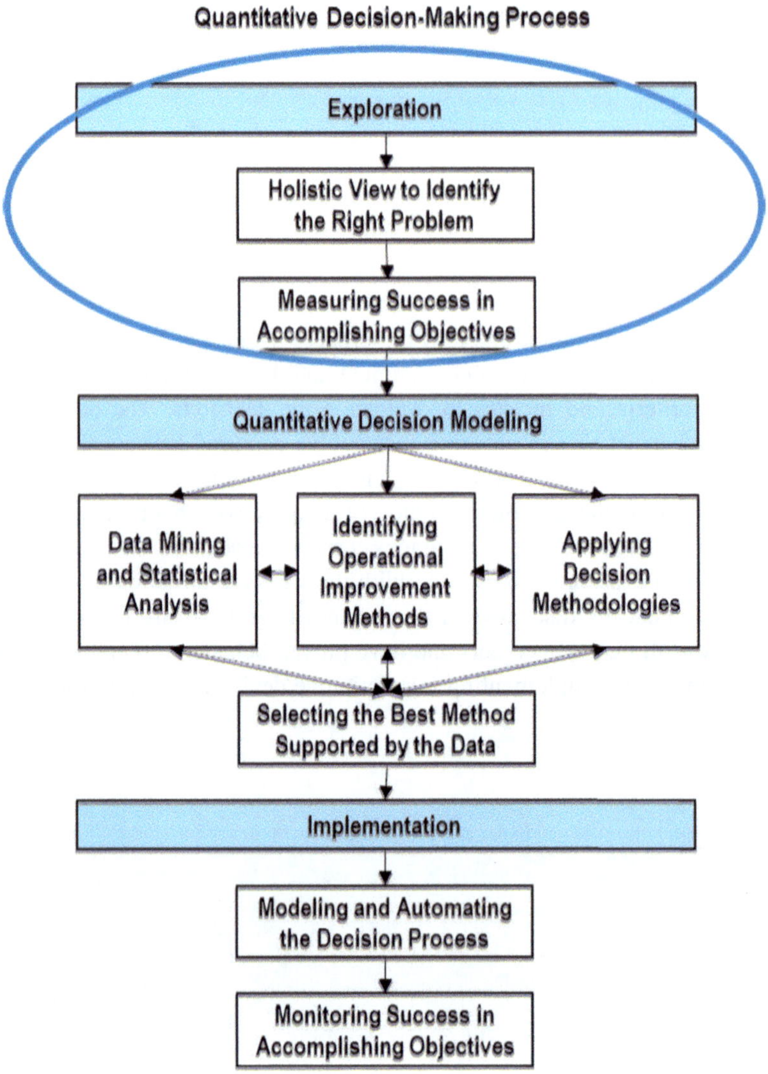

Fig. 1 Exploration

Holistic View to Identify the Right Problem

Abstract

This chapter starts the problem exploration process by discussing the need to gain a holistic view of your operating environment and engage decision makers. Developing a high-level business process model is a starting point. Additionally, asking the right operational questions can lead you to better understanding the issues that your organization faces and the key operational parameters that you should seek to optimize. An example of a bottom-up decision-making process describes potential decision-making complexity. Additionally, resistance to process improvements are discussed to highlight difficulties you may encounter when attempting to make change to improve your organization's operations.

Introduction

Problem solving from a holistic organizational perspective will lead to more effective and efficient operations. Solutions based on an overall objective approach, and not on individual agendas, and a well-operating environment support the future success of a company. Knowing where to begin, what to assess, how to look for workable solutions, and what to measure provides effective solutions that can be implemented by the entire organization.

There are several factors associated with understanding the overall operational environment. You must understand the decision makers, the decision-making complexity, the operating and financial constraints of the environment and have a clear understanding of the cross-functional flow of operations. It is easy to get mired in detail, overlook key elements of the problem, and not know what techniques are most appropriate. Additionally, time is always of the essence, so solutions must be quickly developed to solve problems

D. Cassone, *Operational Leadership Using Quantitative Decision Making*,
Synthesis Lectures on Engineering, Science, and Technology,
https://doi.org/10.1007/978-3-031-67565-2_2

in an ever-changing environment. Many times, you look at the problem in a stove pipe manner and try to solve what appear to be key issues from a singular perspective. You must take a global, holistic approach to understand the overall impact a problem has on the entire organization.

Understand the Operating Environment

To understand the problem, you must understand the operating environment and the decisions that must be made at each level of the organization. This includes defining the questions so that the answers are consistent with the corporate and individual goals. The answers to the following questions should support overall corporate goals.

- How much should I produce of a product in each of the product lines?
- What research items should be funded?
- What are my warehousing requirements?
- How does my production affect my inventory requirements?
- What should I plan to produce to meet customer demand?
- How can I optimize scheduling delivery vehicles?
- And, many others.

You must also determine what you can actually change and what you cannot change within the organization. For example, if an organization wants to determine products to keep in a product line, it is not likely that the flagship items should be removed from the product line without considering their impact on all products.

Fixed operating constraints that you cannot change may also exist. This may include any number of constraints on current operations such as plant capacities, warehouse space, and the amount of resources available to spend in a given year. Usually, a whole range of operating constraints exists from manufacturing, warehousing, distribution, and other areas of the company that affect the decision latitude that can be made.

You need to define the operating constraints and the variables associated with the actual decisions to be made, such as things that can be rationally changed in the normal course of business operations. This may include things such as plant throughput, funding of programs, sales commission structures, and others. The idea here is to determine what decisions can be made or changed and what cannot be changed. Understanding the decision limitations within an organization defines the operating constraints of the decision process [1, 2].

Holistically Modeling the Operating Environment

To understand the problem that is being solved, you must understand the environment in which it is being solved. No activity stands alone in nature, and the inputs, outputs, and interactions between the activity are critical to document. A high-level business process model provides a roadmap for analyzing the company activities and understanding the interactions between various organizational functions. You may, however, need to decompose the high-level view to additional levels of detail to understand the issues you are facing in more detail. When possible, start with a 20–30 block diagram and see if it provides you with a view of the functional interactions. Otherwise, you may need to perform a more comprehensive breakdown of activities to fully understand your issues.

Unfortunately, you may dread "the business process modeling" sessions. The problem with business process modeling sessions is that many times, those performing the activity are so engrossed with the actual process that they lose sight of the purpose of the effort. Keep focused on the effort and avoid non-value add details. The author has seen real benefits in developing an organizational level business process model. Swim-lane models can be helpful to understand the various functions performed by the business and how they interact with each other.

The hope is that a well-developed high-level business process model will develop insights to understanding key issues within an organization. At times, that may not provide enough detail to truly understand key issues impacting your organization. While working at a national telecommunications company, the author co-led an effort to improve device quality. This effort began with developing an end-to-end device launch process mapping activity encompassing corporate activities from product conceptualization through development, market launch and delivery of product to customers. It was an extensive effort that brought together all key functions in the organization to develop a process flow that mapped out each step of the device launch process. This effort took several months to complete, however, provided a roadmap to identify key areas that needed to be improved in the organization. This laborious process documented actions, inputs, upstream and downstream processes across the organization. Numerous critical process improvement areas were identified by this effort. For many years after the effort was completed, this process flow diagram has been used and modified to provide a roadmap for operations. The time on the activity was very well spent but it did require sign-off from the key corporate decision makers to enable their staff to participate in the effort.

Understand the Decision Makers

Ultimately, organizations are run by individuals who have a management structure that reports through their functional areas of the organization to the CEO of the corporation. People are the ultimate decision-makers, so the functional issues along with the individuals who are running the functions must be integrated together to form the basis for operational improvement and decision-making.

For example, you may encounter issues with meeting customer demand and believe that your production line needs to be modernized. Who is impacted by this problem and who needs to be involved in the decision process to provide the resources for this modernization? Fig. 1 shows a chart of who may need to be involved in the decision.

Initially, the production, distribution, sales and marketing managers may be aware of the issue and are either impacted by this lack of meeting sales demand or inability to keep up with production. These individuals are functioning at an operational level and most likely have direct experience with the issues and challenges associated with the performance issues. They may talk or fight among themselves because they are not able to meet their target numbers. These individuals most likely have insight and can provide suggestions from a bottom-up perspective with ideas to be brought forward in their management chain. Who then in their management needs to be involved in the decision-making process?

At this point, as a manager, you could discuss this with your director. Your director may look to engage directors in other functions, and they meet and discuss the issues and potential solutions. Or, based on your organization, you may start a more formal

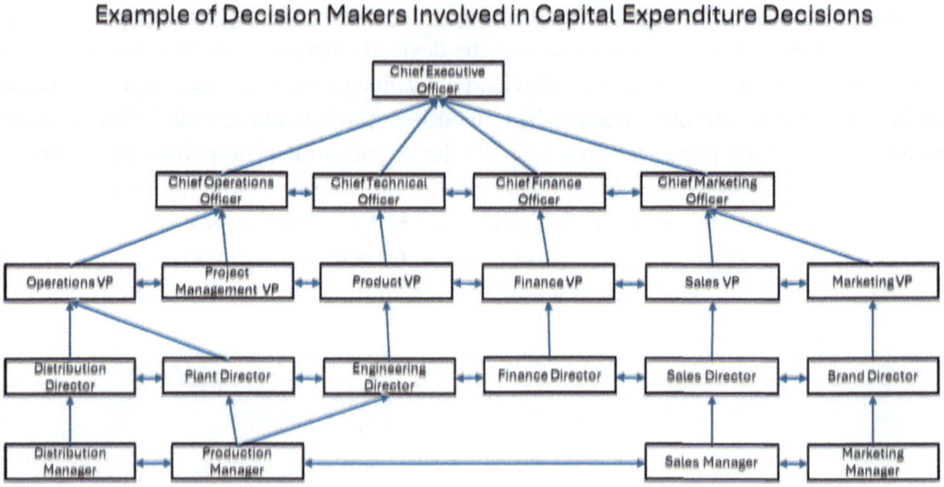

Fig. 1 Potential decision makers involved in a capital expenditure decision

recommendation process, which ultimately will be discussed and will require support at a cross-functional level. Finance would need to be brought in at this point to develop a business case for the modernization.

At the next level of the organization, you are typically providing a formalized request for expenditures and each of the identified management chains at the vice president level would or should be involved. As this is not formally accepted as a project at this point, the Vice President of Project Management or the Project Management Office may be engaged to determine the project management requirements necessary to complete the modernization. Once the more formalized capital expenditure request is developed and approved at the Vice President level, this most likely would then be sent to the senior officers and CEO for approval, based on the capital expenditure oversight threshold of the individual organization.

This is a simplified example of the decision makers engaged in the process from a bottom-up recommendation perspective but does show that decisions require the engagement of multiple functional areas and the engagement of numerous individuals in the management chain. These types of requests may be iterative and driven top-down vs bottom up as described here. When you look at the number of people involved in the decision-making process, you wonder how organizations ever get things accomplished!

The point of this description is for you to assess the organizational requirements in making key decisions. Understanding who your key decision makers are and engaging them in the decision-making process is critical to moving forward ideas in your organization. Sound quantitative methods can be used to improve operations and save money, but the individuals that control those functions should be integrated into the quantitative decision-making process. Explicitly understanding this and purposefully engaging key stakeholders and decision makers in the decision-making process is critical.

Decision-Making Complexity

The process of organizational decision making is complex because executives have their own goals independent of the corporate goals. The challenge is to balance the individual objectives and goals with the corporate objectives and goals both in the short and long term.

Traditionally, decisions are made in a stovepipe fashion. That is, each functional area makes decisions that are best for it without regard to the needs of the other functions; for example, manufacturing is not concerned with marketing or distribution or finance. The reason for this is the reward system for each functional area focuses only on the efficiency for that system. That is, the plant manager or operations manager is rewarded for such things as utilization of assets, return on assets, throughput, quality, minimum labor cost, and scrap for the units produced. None of these factors measure marketing efficiencies or any other functions. Marketing, for example, is basically measured and rewarded for total sales and not on specific products or product mix. When each function goes its separate way and decisions are made that improve that single function, who is looking out for the company as a whole? It begs the question, "Who is running the company?" Is it running by default where the strongest personality drives the company from their functional perspective? How, then, can you develop an overall plan for the good of the company where individual functions are sacrificed a little for the overall good? This may result in manufacturing making products that don't fully utilize the production assets; where marketing doesn't maximize the total volume sales but sells an optimum mix of products that maximize customer service and maximize profitability. You can then see it would be much better overall if there were a combined reward system so that the overall good of the organization is achieved in place of individual goals in the short and long term.

This leads to the issue of how best to manage the many objectives and how to trade-off between them so that the entire organization prospers now and in the future. To do this you must first state the organization's goals and objectives. You must first specifically determine what these goals are in the short term and for the future and then set up metrics to measure how well you accomplish them. One way to do this is to ask some difficult questions. What are you trying to accomplish within the company? What are you trying to accomplish at each organizational level? Unfortunately, most firms do not spend time asking these questions, or if they do, they don't implement the answers into their everyday operations. Why does this happen in nearly all the corporations, large and small and at every level? Although everyone's intentions are good in setting corporate objectives, the major problem is that with today's approach to management, the objectives are ill-defined and consequently impossible to implement. How then can you ensure the corporate objectives are implemented throughout the organization, and how can you measure their success? The primary reason corporate objectives are not implemented at the operations level (which is the only place they can make an impact) is that they usually are subservient to individual managers' objectives.

Thus, you must ask, "How can we get there if we don't know where we are going?" More specifically, what are you trying to accomplish at the corporate level, at the functional level, and most important at the individual or action level? A great deal of effort goes into establishing a 3-year plan and identifying a number of corporate objectives with everyone pretending to agree, knowing fully well the plan will be put on the shelf and promptly forgotten after the exercise is over.

Answering these questions requires getting key individuals to look closely at what they each want to accomplish individually and companywide. This process takes thought and reflection by management. Starting this type of thought process makes management focus on its individual needs and then integrate them into the overall needs of the organization. Many times, individuals are not overtly aware of how solving their problems impacts the other divisions within their company. Bringing these individuals together and addressing the problems forces them to look at their problems as total organizational problems [1, 2].

Difficulties in Implementing Process Improvements

Upper management attitudes can drive operational behaviour in identifying certain problems and developing potential solutions. You must be aware that you can expose issues that may put upper management in a less than favourable light to the organization. This can be dangerous to their career, and if not recognized, dangerous to your own career! Because of this, you must understand the political corporate environment enough to know where and who the potential solutions may impact. Not to say that you should not approach problem solving from a completely analytical and objective position, but you must also know how to present the results in a way that has the best chance of being implemented. This knowledge must guide the problem solver to gain acceptance of the actual solutions by management.

Process improvement efforts identify ways to reduce expenses, improve operations, enhance productivity, speed up delivery and improve customer satisfaction. However, process and productivity improvement efforts be difficult to implement in the organizational culture. Being aware of reasons for resistance can assist your team in understanding how to address these issues and implement improvements. A strong team, good communication, audits of results and an iterative, engaging process can assist in implementing improvements. A few reasons that implementing improvements may be difficult are listed below [3].

- Individuals are territorial and do not want you involved in their business area
- Lack of confidence that the solution will improve the situation
- Failure in previous process improvement efforts.
- Lack of buy-in from senior executives
- Not bought-in on the solution or a lack of trust for the individuals that

Additionally, you must understand the financial climate of an organization. Are funds available for major or even minor changes? Is the industry in a downturn? Has inflation impacted product costs? Are you having difficulty obtaining financing as an organization? If so, projects may or may not be initiated for system enhancements, construction projects, or new marketing campaigns. If funds are not available, implementable solutions may need to be tempered so that they are low- or no-cost solutions. This adds an additional layer of complexity to the problem solver and the solutions generated.

References

1. Cassone, Deandra T., Tillman, Frank A., *A Professional's Guide to Problem Solving with Decision Science: An Integrated Approach for Assessing Issues, Finding Solutions, and Reaching Corporate Objectives*, Second Edition Overland Park, KS, Pioneering Partnerships, LLC, January 2018

2. Tillman, Frank A., Cassone, Deandra T., *A Professional's Guide to Decision Science and Problem Solving: An Integrated Approach for Assessing Issues, Finding Solutions, and Reaching Corporate Objectives*, Upper Saddle River, NJ, Financial Times Press/Pearson Education, Web: December 1, 2012, Hardcover: February 2012
3. Lucidchart Blog, How to Effectively Implement Business Process Change, link: How to Implement Business Process Changel Lucidchart Blog

Measuring Success in Accomplishing the Objectives

Abstract

This chapter discusses the importance of clearly articulating and documenting the objectives of a corporation. Each functions within a company has different objectives with differing importance. Many of the high-level objectives of the corporation may conflict. Senior management should articulate these objectives and reach an overall consensus of the objectives and their importance along with measuring the success of meeting them. This chapter also discusses the development of objectives, decision criteria and metrics which are specific quantifiable measures to determine whether the objectives are met. You should also capture the importance of these decision criteria and metrics since they too may not be equally important to appropriately measure improvement. These objectives and decision criteria form a roadmap for addressing key issues faced by your organization and a means to measure improvement to these metrics.

Establishing Goal-Driven Objectives, Decision Criteria and Metrics

As mentioned in the previous chapter, improvements to your organization may be driven from a bottom-up perspective and championed by those closest to the problem. Many times, however, decisions as to whether certain activities will be funded are driven by top-down decision from senior leadership. For executive leadership to make these strategic decisions, they should have a structured process to evaluate opportunities and ensure that the resources being spent align with their organizational goals and objectives. Time should be spent at an executive level to develop these organizational goals and apply structured and quantitative methods to make these resource decisions. To do so effectively

D. Cassone, *Operational Leadership Using Quantitative Decision Making*,
Synthesis Lectures on Engineering, Science, and Technology,
https://doi.org/10.1007/978-3-031-67565-2_3

and quantitatively requires the development of a purposeful and structured approach to measuring options against metrics that support corporate goals.

Numerous approaches exist to perform strategic planning and define organizational goals and objectives. The author and Dr. Frank A. Tillman present steps in their book, *A Professional's Guide to Problem Solving with Decision Science* [1, 2]. Overall, you must define what you are trying to accomplish, how you intend to measure that accomplishment and have consensus from senior leaders on what those key objectives and measures are for the organization.

Understanding what is important to your organization is critical and this type of exercise ensures that activities performed, and resources expended align with what is important to your organization. Otherwise, your efforts may not be accepted by your organization or be a sub-optimal representation of your organization's goals. Establishing goals and objectives provides the organization with a definition of its goals and objectives that represent its current thinking. These goals and objectives can then be integrated into the decision-making process so that decisions made with the decision model are structured to impact all levels of the organization and are quantifiable and defensible. With quantitative decision making, you must know what is important to your organization to deliver strategic solutions.

In establishing corporate goals and objectives, it is most effective when executive management is engaged in the process and a consensus is reached among those decision makers as to how you will measure your resource expenditure considerations. A group consensus should be gained from these decision makers. Even if some choose to "agree to disagree" on certain critical objectives and metrics, conducting sessions to gather this information is valuable to leadership and the organization. Available group decision-making techniques vary in degrees of formality, and you can use them to facilitate this process. Techniques range from simple brainstorming to more robust methods such as the Nominal Group Technique (NGT) or Sequential Proportional Additive Numeration (SPAN) methods. The end result should be a clear definition of what a corporation views as its goals and objectives. When defined, these goals and objectives can then be built into the decision process. The following is a sequence of steps that you can use to establish the goals and objectives of an organization.

This approach to metrics development has been used in many private sector and government evaluations. The approach is to relate metrics directly to the accomplishment of the goals and objectives of the organization. This process requires the management team to critically examine goals and objectives to ensure the decisions directly relate to these objectives. Decision makers must first establish the goals and objectives and their relative importance in the final decision process.

Decision criteria and metrics are then established to support these established goals and objectives. Group decision-making techniques are also utilized to provide a means for developing the decision criteria and weighting their importance. All the decision makers should have a say in the final selection decision to ensure objectivity and avoid having dominant personalities overly control the process. The steps involved in this time-tested approach follows:

Step 1 Establish overall objectives and goals.
Step 2 Weight the objectives to determine their importance.
Step 3 Select the decision criteria.
Step 4 Weight the criteria to determine their importance.
Step 5 Develop metrics.

Overall, this approach provides a consistent, traceable, and defensible basis for making decisions.

Step 1: Establish Goals and Objectives

The first step involves establishing the objectives for the organization. Primary decision makers should meet together to ensure that all the components of the organization are represented in the decision process and are represented in the objectives. One approach is to provide to the group a "straw man" list of corporate objectives as a starting point in the development of the final list of objectives. The straw man list is based on the objectives that have historically been deemed important to the group and may include objectives such as maximizing profitability, minimizing risk, minimizing cost, or maximizing growth. The management group can then use these as a starting point and brainstorm to add or delete from the list. As part of this process, the group should also provide a high-level definition of the objectives so that all involved understand what is meant by them.

Managers may not have the authority to affect high-level decisions, however, the establishment of these objectives and the development of the subsequent decision criteria and metrics give managers a defensible position in their decisions. The objectives established in this manner can provide the building blocks for establishing the decision criteria and metrics used within the decision model.

It is also beneficial to establish a mission statement for the organization. A mission statement is a succinct statement, typically one sentence that summarizes the purpose of the organization. A mission statement describes why the organization is in business and what it wants to accomplish.

When any group of individuals meet to conduct activities such as this, it is difficult to arrive at complete agreement. You need to understand upfront that this may be the case. In this process, you must leave the latitude to "agree to disagree." That is, it is okay if there is not complete agreement on all the objectives. Not all objectives may have the same importance to each function of an organization. Weighting the objectives, as described in the next section, provides a means to address that not all individuals may feel the same about the importance of each objective, which can be handled in the modeling process.

These objectives are broad in nature and can focus management on organizational-level goals. Objectives are weighted to further focus corporate management on key operating

Table 1 Example of goals and objectives for a company

Objectives	Definitions
Improve financial position	Improve the financial position by evaluating the project or product's financial contribution to the company
Enhance technology development	Improve the technological position by evaluating the project or product's ability to meet the technology development and innovation needs of the company
Improve market position	Improve the market position of the company with this product or project

areas within the company. Specific decision criteria and metrics further refine the key attributes that constitute the corporate measurement of the objectives. The objectives provide the starting point for the decomposition of strategic goals into actionable activities and problem solving.

Table 1 shows an example of corporate goals and provides a high-level definition of these metrics.

Step 2: Weight the Objectives to Determine Their Importance

Many times goals and objectives can be stated as being important to an organization but the relative importance of them is not quantified. In addition, there are typically conflicting objectives in which the improvement in one area may adversely impact another area. For example, improvements in information technology may also lead to increases in capital costs.

This step involves generating an overall importance for each of the established goals and objectives. This is accomplished by having each of the team members weigh the objectives established in the previous step for their view of overall importance in decisions. You can simply average the weights provided by the individuals on the team or can use other decision science methods. An example of objective weighting is shown in Table 2.

Consensus Building

Consensus building is fundamental to group decision making. The problem of group decision making can be broadly classified into two categories in this field: expert judgment and group consensus. The expert judgment process entails making a decision by using expertise gained through experience. Specifically, it is concerned with making judgments and constructing new solutions to the problem. On the other hand, the group consensus process involves groups that have common interests, such as executive boards and organizations, making a decision.

Table 2 Example goal weighting

Weights (%)	Objectives	Definitions
40	Improve Financial Position	Improve the financial position by evaluating the project or product's financial contribution to the company
20	Enhance Technology Development	Improve the technological position by evaluating the project or product's ability to meet the technology development and innovation needs of the company
40	Improve Market Position	Improve the market position of the company with this product or project

Table 3 Example of consensus weighting for objectives

	Decision maker weights					
Objectives	1 (%)	2 (%)	3 (%)	4 (%)	5 (%)	Consensus
Improve Financial Position	40	50	40	60	30	44
Enhance Technology Development	30	20	50	30	40	34
Improve Market Position	30	30	10	10	30	22

Expert judgment and group consensus methods involve methods commonly utilized in the group decision-making process. Creative methods used to extract, generate, and stimulate new ideas may include brainstorming, brainwriting or Nominal Group technique. Techniques that are used to explore and clarify existing issues might include surveys, conferences, and the SPAN technique. An example of group consensus weighting is shown in Table 3. Additional group decision making techniques are provided in a later chapter of the book.

A weighting scheme can also be generated by the Normalized Direct Weighting scheme where each team member assigns a score on a scale from 1 to 10, where 10 was the most important and average the results to provide a group consensus.

Another approach to weighting objectives is to determine the percent of budget that falls within a given corporate functional area. This provides a view of the potential impacts of improvement opportunities for a given functional area.

Assigning an overall importance weighting to the objectives gives the decision makers an idea of what the group as a whole view as the importance of the objectives for the organization. These objectives and their importance also provide the necessary direction and focus in the decision modeling process. You now have a better idea of what you can accomplish as an organization.

Step 3: Establishing Decision Criteria and Metrics

You need to know what the organization wants to accomplish, the purpose of the decisions being made, the operating environment of the organization and how you can measure the accomplishment of the organizational objectives. You can then develop the evaluation process to reflect the conditions and characteristics of the organization.

When an organization has defined and agreed upon its goals and objectives, decision makers should then develop a means to track and measure the accomplishment of these objectives. Often within an organization, a strategic plan, short-range plan, or long-range plan is developed and published as an edict to the organization. These plans may, however, end up on a shelf with no real meaning in the day-to-day operation of the organization. To drive these goals into the operating decision levels within the organization, these goals and objectives must be translated to meaningful measures of success that are used within the decision process and tracked within the organization. This is key to ensuring that the organizational goals are met. In the last section you saw how to integrate these organizational goals and weight their importance for all to follow. You now need to track their success.

Because so much data is currently available, you need to determine those measures (metrics) that provide a meaningful measure in achieving success within the organization. This process involves identifying criteria that are important to the organization. Typically, decisions are made based on one or two criteria, so the metrics to support the decision process should be kept to a minimum. An approach is described in this chapter for identifying decision criteria and determining their overall importance in the decision process and measuring their success.

Based on identifying the important criteria, data sources are assessed to determine whether this information can be relatively easily obtained, processed, and maintained. Data should be assessed to determine what level of detail should be maintained for it to be used as a metric to measure success. It may be tracked, for example, on a plant-by-plant basis or rolled up across the organization as a whole. Data can come from both automated sources or from assessments made by experienced executives or experts in the field. The process to establish, track, and maintain metrics and measures of success for an organization provides meaningful information in achieving success and understanding the organization's operations at every level.

Step 4: Select the Decision Criteria

For each of the objectives established in step 1, the group must establish a hierarchy of decision criteria to represent the various objectives. Define the decision criteria so that there is a clear understanding of the criteria used and what metrics will be used to measure the criteria. This definition phase provides the framework for establishing the

Table 4 Example of decision criteria supporting corporate objectives

Objectives	Decision criteria
Improve financial position	Return on resources invested
	Impact on operating cost
	Net present value
Enhance technology development	Probability of technological success
	Technology development requirements
	Innovation
Improve market position	Market Size
	Market growth potential
	Market attractiveness

metrics associated with each of the decision criteria. A "first cut" of the overall decision criteria will be developed, reviewed, and revised as necessary by the decision makers and senior management. Table 4 shows an example of the corporate goals with their associated decision criteria.

Group consensus building techniques will be required to arrive at the decision criteria that support the corporate goals and objectives. It may also be difficult to narrow down those key decision criteria that support accomplishing corporate objectives. However, the discussion and understanding that is developed in this process can be very beneficial to the corporate participants and leaders. The end result will be a focused understanding of what areas of measure are important to the company's success.

The decision criteria should then be defined to ensure that all corporate participants have an accurate understanding of the meaning of the decision criteria. Additionally, the units of measure should be agreed upon by the group. This involves determining the type of data that will be used to evaluate the project or product (subjective data, objective data) and any scaling associated with the data in the analysis. This provides a clear framework for evaluation for the corporate participants. Table 5 shows an example of the decision criteria definitions that support corporate objectives.

Step 5: Weight the Criteria to Determine Their Importance

Team members then weight the criteria established in step 3 for their relative importance in the decision process. Again, use group decision-making techniques to facilitate this process. The team must develop a "first cut" of the decision criteria weights and review and revise their findings to ensure reasonableness.

Table 6 shows an example of the decision criteria weighting for each of the team members. Their importance score is then averaged as a consensus weighting for the group.

Table 5 Example of decision criteria supporting corporate objectives with definitions

Objectives	Decision criteria	Definitions	Objective or subjective data for metrics
Improve financial position	Return on resources invested	Projected return on the resources invested in the product or project	Objective
	Impact on operating cost	An assessment of how this product or project will impact operating costs. Scaled assessment of negative impact (increase), neutral or positive impact (reduce)	Subjective
	Net present value	The Net Present Value of the product or project using the company's discount rate	Objective
Enhance technology development	Probability of technological success	An assessment of the probability of the successful development of this technology	Subjective
	Technology development requirements	An assessment of whether the development of this technology requires new products or processes or whether it can be accomplished with currently available capital assets or technologies	Subjective
	Innovation	Is this an innovative technology, enhancement to current technology or are similar to currently available technologies	Subjective
Improve market position	Market size	The projected market size of this product or project	Subjective

(continued)

Table 5 (continued)

Objectives	Decision criteria	Definitions	Objective or subjective data for metrics
	Market growth potential	The potential impact that this product or project has on market growth	Subjective
	Market attractiveness	The market attractiveness of this product or project. This is a subjective assessment but should be based on market research	Subjective

Table 6 Example of consensus weighting for decision criteria

		Decision maker weights					
Objectives	Decision Criteria	1 (%)	2 (%)	3 (%)	4 (%)	5 (%)	Consensus (%)
Improve financial position	Return on resources invested	30	20	25	40	15	26
	Impact on operating cost	40	60	40	30	35	41
	Net present value	30	20	35	30	50	33
Enhance technology development	Probability of technological success	45	30	35	45	40	39
	Technology development requirements	30	40	25	25	35	31
	Innovation	25	30	40	30	25	30
Improve market position	Market size	20	30	35	40	35	32
	Market growth potential	50	25	35	30	35	35
	Market attractiveness	30	45	30	30	30	33

Table 7 shows an example of the weights resulting from the consensus objective weights and the corporate decision criteria weights. The resulting criteria weights show how important the team members viewed each of the individual decision criteria. This information will be used further in our evaluation process.

Table 7 Example of weights resulting from the objective and decision criteria consensus weighting

Objective weights (%)	Objectives	Decision criteria weights (%)	Decision criteria	Resulting weight (%)
40	Improve financial position	26	Return on resources invested	10
		41	Impact on operating cost	16
		33	Net present value	13
20	Enhance technology development	39	Probability of technological success	8
		31	Technology development requirements	6
		30	Innovation	6
40	Improve market position	32	Market size	13
		35	Market growth potential	14
		33	Market attractiveness	13

Step 6: Develop Decision Criteria Metrics

From the decision criteria established in step 3, you can identify the metrics. This involves determining what data to use to measure and quantify the decision criteria. The criteria can be either subjective or quantitative in nature. You can measure criteria using "hard," quantitative data or a subjective scale of the decision makers. Expert opinion can be subjectively used when objective data is not available or when objective data is too costly or time-consuming to obtain. Again, the group must develop a "first cut" of the decision criteria metrics and review and revise its findings as needed to satisfy the decision makers.

Importance of Metrics

A metric is a standard measure to assess performance in a particular activity. A metric is a composite of measures that yield systematic insight into the state of the process or products and drives appropriate action. Metrics can be composed of both objective data such as historical data reported in a database and subjective data such as expert opinions by senior management or experts in the business.

Metrics are important for a number of reasons. You can use them to defend and justify decisions, to provide objective assessment of progress toward goals, and for problem solving and to validate process improvement. Successful enterprises constantly assess themselves and improve in all dimensions of their enterprise. Metrics provide a foundation for the assessment of success of an activity or enterprise.

A good metric is built upon an organization's missions, goals, and objectives. A metric must be meaningful and understandable for management and workers alike. The best metrics derive naturally from the process in which the data is relevant to the process and its collection becomes part of the process. The metric itself must be easily measurable. The requirement for new systems and data to implement a metric should be minimal because if not, data will not be gathered to support it.

Table 8 shows an example of decision criteria and whether these criteria will be quantified with objective or subjective measures and the measures to be used in the evaluation.

You can use either objective or subjective data to represent metrics used in the decision process. Objective data usually can be described as data that can be quantified by some measure of known commonality. This may be data such as the number of items produced, number of trucks in a location, population of a city, and so on. This data is usually available in some form in company databases and information systems. Typically, statistics such as averages and trends are generated based on a record of this objective data over some point in time. Objective or quantitative data represents a history of activities of a company that has been operating during a given time period.

Qualitative or subjective data can be easily used in a number of different situations. Surveys are good examples of subjective data used to represent a rating of a product or service. Use scales from 1 to 5 or 1 to 10 to represent high, medium, and low assessments for a given metrics. Use assessments such as red, yellow, and green in other situations in which individuals (such as military personnel) might find more meaning in rating conditions. Numerical values with their verbal description provide the type of information that can be captured and utilized in a decision model when other information is not available. Subjective data is data based on someone's opinion or best guess of a condition or a future event.

In developing goals, decision criteria, and constraints, consider several parameters in the development process to ensure a set of well-structured, well-represented goals and decision criteria. These development parameters are as follows:

- Goals and decision criteria must represent actual and important considerations in making decisions.
- Decision criteria must differentiate one project from another in terms of higher or lower priority.
- Decision criteria must be independent, not overlapping in content or intent, to avoid accounting for the same thought or idea more than once.
- Decision criteria must be defined as clearly as possible to ensure that the decision criteria in the evaluation process are viewed in the same context.

Table 8 Example of decision criteria definitions and metrics

Objectives	Decision criteria	Definitions	Objective or subjective data for metrics
Improve financial position	Return on resources invested	Projected return on the resources invested in the product or project	Objective—Percentage
	Impact on Operating Cost	An assessment of how this product or project will impact operating costs. Scaled assessment of negative impact (increase), neutral or positive impact (reduce)	Subjective—Based on projected operating costs
	Net present value	The Net Present Value of the product or project using the company's discount rate	Objective—Dollars
Enhance technology development	Probability of technological success	An assessment of the probability of the successful development of this technology	Subjective—Percentage
	Technology development requirements	An assessment of whether the development of this technology requires new products or processes or whether it can be accomplished with currently available capital assets or technologies	Subjective—Subjective scale addressing the ability to leverage existing technology or the requirement = 5 to develop new resources to support = 1
	Innovation	Is this an innovative technology, enhancement to current technology or are similar to currently available technologies	Subjective—Subjective scale where technology is innovative = 5 or similar to available technologies = 1
Improve market position	Market size	The projected market size of this product or project	Subjective—Scaled from Low = $5 million to High = $100 million

(continued)

Table 8 (continued)

Objectives	Decision criteria	Definitions	Objective or subjective data for metrics
	Market growth potential	The potential impact that this product or project has on market growth	Subjective—Scaled from Flat/Negative = 1 to > 20% = 5
	Market attractiveness	The market attractiveness of this product or project. This is a subjective assessment but should be based on market research	Subjective—Scaled from Low = 1 to High = 5

- Measures and scales developed for the decision criteria must be meaningful in the evaluation process and the data to perform the evaluation easily accessible.
- Constraints that represent types of mixes, qualifiers, and conditions that would be applied to a prioritized list of items must be identified and differentiated from the evaluation decision criteria.

In quantitative decision making, it is important to conduct an exercise such as this to quantify decisions and measures of performance and success in your operations. Although the process discussed here is simple, the discussion, interaction and consensus building activity with the executives can be very challenging. You may choose to use other methods to accomplish this, but it is recommended that you use some similar process so that you can quantify the impacts of your decisions [1, 2].

References

1. Chapter adapted from: Cassone, Deandra T., Tillman, Frank A., *A Professional's Guide to Problem Solving with Decision Science: An Integrated Approach for Assessing Issues, Finding Solutions, and Reaching Corporate Objectives,* Second Edition Overland Park, KS, Pioneering Partnerships, LLC, January 2018.
2. Chapter adapted from: Tillman, Frank A., Cassone, Deandra T., *A Professional's Guide to Decision Science and Problem Solving: An Integrated Approach for Assessing Issues, Finding Solutions, and Reaching Corporate Objectives,* Upper Saddle River, NJ, Financial Times Press/ Pearson Education, Web: December 1, 2012, Hardcover: February 2012.

Quantitative Decision Modeling

Part II addresses key building blocks that can be used to develop quantitative decision models. Data mining and statistical analysis, operational improvement methods and quantitative models and applying decision models all work together to develop a methodology that best supports quantitative decision making. (See Fig. 1.)

Understanding what your data is saying to you is critical. Knowing robust quantitative methods that can be used to model your current issues is important to implement solid solutions. Ensuring that your executive decision makers are engaged in the decision process provides a means for robust approaches to measuring and monitoring solutions. All of these components are integrated together to develop improvement methods that are best supported by the data and the decision makers.

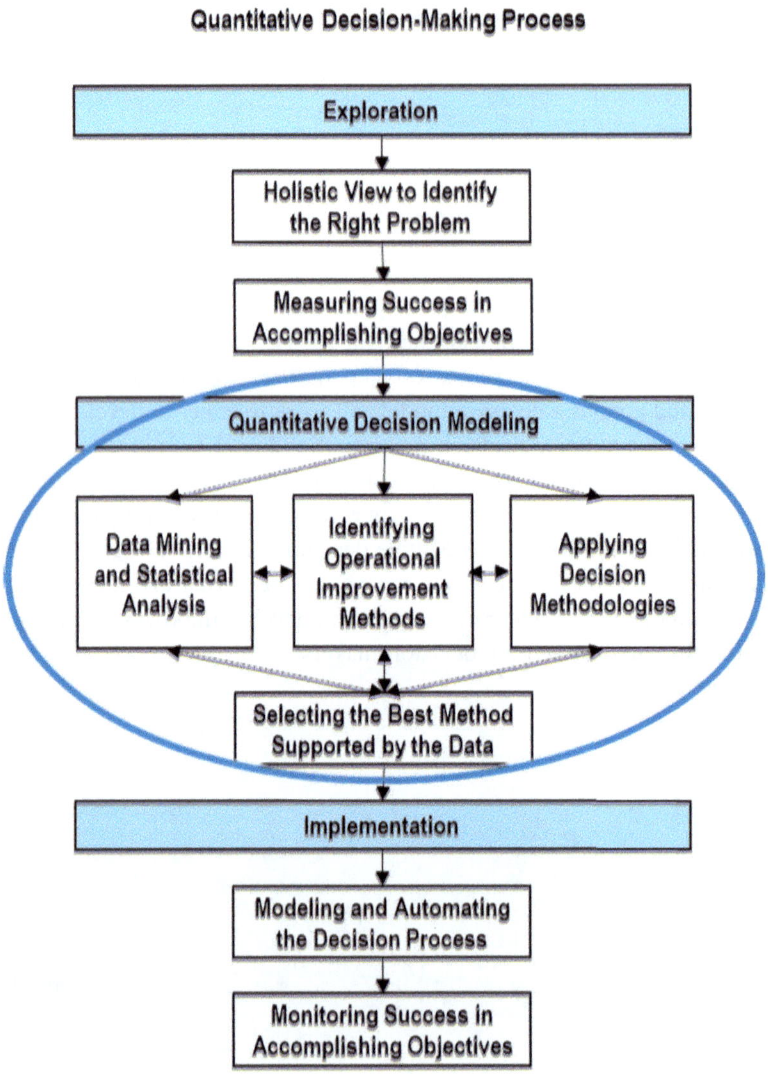

Fig. 1 Quantitative Decision Modeling

Data Mining and Statistical Analysis (Data Analysis and Performance Evaluation)

Abstract

Data mining and statistical analysis forms a foundation for understanding how your organization is actually operating. Simple, descriptive statistics can provide great insight into your organization's operations and can be easily calculated. Diagnostics analytics help you understand the root cause and relationships associated with your operations. Predictive analytics assist you with forecasting the future. Prescriptive analytics employ more sophisticated techniques that can assist you with data-driven decisions. Data and data analytics are only one part of successful operational leadership with quantitative decision making but form a foundation in understanding the current and future state of your operations.

Fundamental Data Analysis

Data provides insight into the operational aspects of your environment. So much data is currently available, it is often difficult to sort through the stacks of data to determine what is or is not important. Often data analysis is all that you need to solve the problem. Other times you can use data and statistical analysis in the analysis and modeling process.

Spreadsheets and database languages are relatively easy to use, and you can use them to perform numerous analyses on available data within a short amount of time. The ability to graph data as well can provide insight into trends and performance of the corporation. Some tools such as Microsoft Excel have built-in statistical analysis functions that can aid to quickly analyze data and information. You can use forecasting to predict future activities based on the historical data. To predict the future, the use of expert opinion and the judgment of those experts in the field can be one of the best predictors. With these readily available tools to analyze your data, there really is no reason not to look

D. Cassone, *Operational Leadership Using Quantitative Decision Making*,
Synthesis Lectures on Engineering, Science, and Technology,
https://doi.org/10.1007/978-3-031-67565-2_4

at the data that you have available to gain insights into the information and operation. Artificial intelligence and Neural Networks, provides a statistical basis for forecasting future direction. These approaches require large amounts of data but are useful in making future predictions. Key in data analysis is to determine what the information can tell you about the operation.

The complexity of the business environment, the objectives to be accomplished by the organization, the availability of data, and the time frame for the development of the model can dictate what methods and analysis you should use to solve the problem. Quantitative decision-making models may require combining a number of methods, such as multiple objective decision making, statistical analysis, simulation, and expert system technology. Don't be afraid to combine these methods and models to arrive at a solution. Many times, you can't solve a problem with just one method. Allowing yourself the flexibility to integrate multiple methods and models, even if it has never been done before, may lead you to the best representation of your operating environment. This may also lead to business process innovation.

The decision maker must understand how the quantitative decision process works to become comfortable with the decisions generated because his future and the future of the company depends upon the decision to be made. Keep the decision-making methods understandable and explainable to the decision maker. Thus, the simpler the method selected, the better. The best method is the one that the decision maker feels comfortable with and uses.

Integrating the data analysis, the solution process, together with the decision maker goals, ensures the decision-making model can support the decisions to be made. The models developed should be automated to the degree that is feasible. An automated model provides the ability to quickly perform sensitivity analysis. Generating the first answer from a model is only the start of the analysis for decision making. You need to test various parameters and scenarios in the decision process to determine the best solution. Updating the data can keep the model current and representative of the changing business operations. As the business climate changes, the model should be dynamic and adapt to the changing environment.

There is no predetermined analysis that should be performed on data. Data analysis is based on understanding the operating environment, objectives of the model and the availability of the data and the decisions to be made with the model. You need an understanding of the problem before you can perform any beneficial data analysis. Make a number of queries and analyses and scrutinize the data to gain an understanding of the problem and accuracy of the data. One analysis leads to other analyses, which may lead to a direction you had not originally planned. Honing your basic skills in statistical analysis can greatly benefit you in understanding and interpreting your operating environment.

As the world of big data analytics has evolved, data analytics has been categorized into four basic categories. Additionally, methods associated with these classifications continue to evolve and it is worth keeping abreast of new developments as they may be useful in solving problems. There are four main types of data analytics which are briefly described below [1].

- Descriptive analytics—This type analytics focuses on summarizing historical data. Many of the simple statistical techniques such as means, modes, etc. fall into this category along with understand trends, patterns, and key performance indicators (KPIs).
- Diagnostic analytics—This type of analytics focuses on identifying the root causes and relationships associated with the data, such as that found in regression and correlation analysis.
- Predictive analytics—This type of analytics uses historical data to predict future events such as you might find using forecasting techniques.
- Prescriptive analytics—This type of analytics builds on predictive analytics to recommend future actions and to make data-driven decisions and require neural networks or artificial intelligence.

Descriptive Analytics

This type of analytics focuses on summarizing historical data. Many of the simple statistical techniques such as means, modes, etc. fall into this category along with understand trends, patterns, and key performance indicators (KPIs). This type of analysis can provide you with useful information that can help you understand your operating environment.

Standard Statistics

The following are some standard statistics that you can use to analyze current and historical data. These statistical functions are readily available in Microsoft Excel or other standard tools. You should look at these basic statistical values when assessing your data. These analyses are easy to do and can provide great insights into your operations with a small amount of effort.

- Mean—The arithmetic mean or average of the data set.
- Low—The lowest value in the data set.
- High—The highest value in the data set.
- Median—The value that represents the middle of the data set when ordered. If there are not equal occurrences above and below the median, an average of the two middle values is taken.
- Mode—The most frequent value in the set of data.

- Standard deviation—The measure of dispersion of a frequency distribution that is the square root of the arithmetic mean of the squares of the deviation of each of the class frequencies from the arithmetic mean of the frequency distribution.
- Variance—The measure of dispersion of a frequency distribution that is the arithmetic mean of the squares of the deviation of each of the class frequencies from the arithmetic mean of the frequency distribution.

Plotting your data to gain a view of the patterns and information that it presents is also helpful. This helps you to understand the dispersion of data and see any trends that may lead you to further analysis. Histograms and frequency distributions by deciles can also provide additional insight and are described below.

Histogram

A histogram is a representation of a frequency distribution divided by classes of data and plotted according to the frequency of the occurrence for each of the classes of data against another axis, for example, time, color, and so on. Most likely, the data used to develop the histogram will be represented by integer groupings, that is, 1 day or 2 days, or can be grouped into ranges of data such as 0–3 days, 4–6 days, and so on. You can then see how often your data falls into each of these categories. This plot then shows the This plot can be performed using a line graph or with bar graphs, whichever is the typical representation for this data.

Frequency Distribution by Deciles

This is similar to the development of the histogram; however, the frequency distribution would be divided into 10 equal classes of data such that each class contains the same number of individual data points. Ranges or averages associated with each of the classes of data would display on the plot of the information and can also display in tabular form.

Diagnostic Analytics

This type of analytics focuses on identifying the root causes and relationships associated with the data, such as that found in regression and correlation analysis. This type of analysis provides insights into the relationship between variables and the strength of those relationships.

Regression Analysis

This method involves determining the relationships between the dependent and independent variables and representing this relationship in a regression equation. The results of this statistical analysis can be shown in an equation, in tables, or by plotting the regression line for the data set. Regression is a functional relationship between two or more correlated variables that is often empirically determined from data and is used to predict values of one variable when given values of the others.

Regression equations can be used to predict future requirements or activities based on variables input into the regression equations. For example, variables or drivers can be utilized to predict future workloads based on the past relationships between the workload drivers and the resulting work performed.

You can also use stepwise regression to develop a multiple variable equation for forecasting. Do this by adding new variables to the equation and checking to see if the new equation provides a better forecast; for example, reduce the variability around the forecasted value. Stepwise regression uses a statistical test called the F-Test to identify this subset of variables. The F-Test is a test that validates whether the variance is reduced significantly around the forecasted value. In the regression analysis, the F-Test would test whether the variance of the variables used in the regression equation is equal to or less than the variance in the original data. Based on the F-Test, variables are either entered if they reduce the variance or removed if they don't from the subset one at a time until the optimal combination of variables is found. The resulting subset is the minimal best set of variables that significantly reduce the variance and are the most accurate in the predictions or forecasts. Time series regression is a statistical method for predicting a future response based on history and can help you understand and predict system behavior based on the captured data. Figure 1 shows an example regression analysis generated in Microsoft Excel.

Correlation Analysis

Correlation analysis is used to measure the relationship between two variables. Correlation analysis is a statistical test primarily used to identify and explore linear relationships between two variables and then determine the strength and direction of that relationship. The degree of association in correlation analysis is measured by a correlation coefficient. The Pearson correlation, which is denoted by r, is the most commonly used coefficient. The correlation coefficient quantifies the degree of linear association between two variables and can take values between -1 and + 1. There is no linear correlation between variables when the r value is zero. The direction of the linear relationship between two variables by the correlation coefficient's sign where a positive correlation means that both variables increase simultaneously and a negative correlation means that as one variable

Month	Month	Quantity	Y - Predicted
January	1	1325	1356
February	2	1353	1325
March	3	1305	1293
April	4	1275	1261
May	5	1210	1230
June	6	1195	1198
July	7		1166

SUMMARY OUTPUT

Regression Statistics	
Multiple R	0.935
R Square	0.875
Adjusted R Square	0.844
Standard Error	25.055
Observations	6

ANOVA

	df	*SS*	*MS*	*F*	*Significance F*
Regression	1	17569.729	17569.729	27.987	0.006
Residual	4	2511.105	627.776		
Total	5	20080.833			

	Coefficients	*Standard Error*	*t Stat*	*P-value*	*Lower 95%*	*Upper 95%*	*Lower 95.0%*	*Upper 95.0%*
Intercept	1388.067	23.325	59.509	0.000	1323.305	1452.828	1323.305	1452.828
Month	-31.686	5.989	-5.290	0.006	-48.315	-15.056	-48.315	-15.056

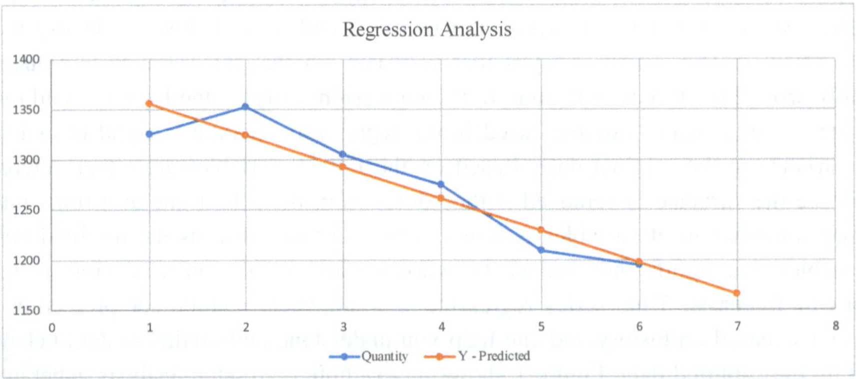

Fig. 1 Example regression analysis generated in microsoft excel

increases, the other variable decreases proportionally. A larger sample size will improve the confidence and accuracy in the results [2].

Predictive Analytics

Predictive analytics use historical data to forecast or predict future events. You can use several different forecasting techniques to predict future occurrences based on historical data. The most common techniques are moving averages, exponential smoothing, and multiple regression analysis. These techniques are discussed in much more detail in numerous statistics books. Forecasting is important to understand information that is contained in the data. Using what has happened in the past as shown in the data, you can see what may occur in the future if the same patterns occur in the future.

Moving Averages

A moving average is a forecast for a future occurrence of an activity based on the most recent occurrences of the activity. The simple moving average is the arithmetic mean of the n most recent observations. The following are several characteristics of this model. First, equal weights are assigned to the most recent n observations. Second, each new estimate is computed by adding the new data point and discarding the oldest data point for the previous nth period. Thus, each new estimate is an updated version of the preceding estimate. Third, the rate of response of the moving average to changes in the underlying data pattern depends upon the number of periods included in the moving average. In general, the more periods included in the computation, the less sensitive it will be to changes in the pattern of the data. Conversely, a small value of n leads to a moving average that responds relatively rapidly to changes and may have much more variability.

Weighted Moving Averages

Creating weighted moving average forecasts is based on utilizing the concepts of the moving average; however, possibly weighting the most recent observations more in the forecasting process. The weighted moving average enables the observations to be weighted such that more importance can be attached to the more recent observations. The weights should sum to one.

Figure 2 is an example of a 3-month, 5-month, 3-month weighted moving average and 5-month weighted moving average.

Exponential Smoothing

In the exponential smoothing method, new forecasts are derived by adjusting forecasts made for previous periods by considering its forecast error. In this way, the forecaster can continually revise the forecast based on past experience. The model has the advantage of a weighted moving average method, in that more recent observations are assigned larger weights. It reacts faster than the moving average model to changes in the variability of the data. Single exponential smoothing is a procedure in which the forecast for the next period equals the forecast for the prior period adjusted by an amount proportional to the

Month	Quantity	3-Month Moving Average	Weighted 3-Month Moving Average	5- Month Moving Average	Weighted 5-Month Moving Average
January	1325				
February	1353				
March	1305				
April	1275				
May	1210				
June	1195				
July		1227	1216	1268	1242

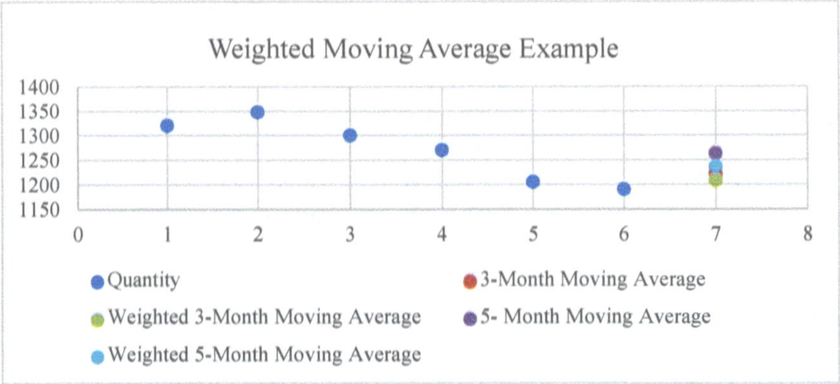

Fig. 2 Moving and weighted moving average examples

most recent forecast. Double exponential smoothing may also be used to address trends in the data.

The smoothing constant must be determined judgmentally, depending on the sensitivity of response required by the model. The smaller the value of, the slower the response. Larger values of cause increasingly quicker reactions in the smoothed (forecast) value. Another difficulty with this method occurs when trying to forecast more than one period ahead. Therefore, this method is designed for and is best to use to forecast only one period ahead. Figure 3 is an example of Exponential Smoothing forecasts.

Expert Opinion

Expert opinion and institutional knowledge can also be captured and incorporated into data models. This knowledge can be outside of your traditional data systems but can be extremely valuable. Expert opinion represents the sum total of information in systems and insights based on experience and should be included in quantitative decision-making

		0.2	0.8
Month	Actual	α = .2	α = .8
January	1325	1370	1370
February	1353	1361	1334
March	1305	1359	1349
April	1275	1349	1314
May	1210	1334	1283
June		1309	1225

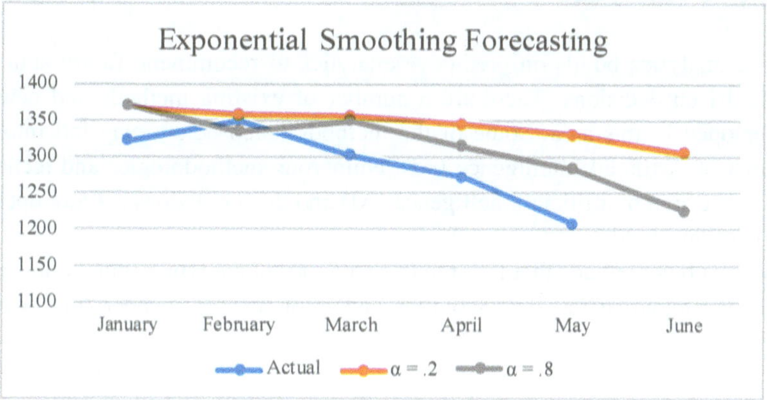

Fig. 3 Example of exponential smoothing forecasts

models whenever possible. Scaling systems can be used to capture thought in a quantitative model in simple scales such as 1–5 or 1–10 with descriptors associated with the numerical value. Expert opinion can be very helpful in truly gaining insights into your organization and operations. Some examples of how Expert Opinion can be captured include the following:

- Judgment and opinion of project success
- Opinions of which are the key areas for strategic planning
- New product potential
- Establishing organizational goals and objectives
- Ranking available options
- Rating the decision criteria used to assess projects
- Providing assessment of future projections

Expert Opinion is often overlooked in the decision modeling process. Expert Opinion integrates the knowledge of an individual and where his or her sum total of experience determines his or her perception of the future. With a structured approach, this knowledge can be quantified, captured, and integrated into the decision process. Much valuable information can be captured from individuals with in depth knowledge about an organization and if not obtained this way would not be used to help make decisions. Expert Opinion is forward looking whereas statistical data, on the other hand, is solely based on backward or historical observations. The assumption in statistical models is that the future will be the same as the past, whereas Expert Opinion can capture what will happen in the future in a fast-changing business climate in the future.

Prescriptive Analytics

Prescriptive analytics builds on predictive analytics to recommend future actions and to make data-driven decisions. There are a number of existing methods and new methods being developed in this area. Some of the methods listed below may fall into the overarching area of Artificial Intelligence (AI). Numerous methodologies and techniques are available in the area of artificial intelligence (AI) and this book does not attempt to capture this ever evolving field.

Below are a few methods that fall into this category of analytics, with many more methods being developed in research activities and implemented in operating environments [1].

Genetic Algorithms

Genetic Algorithms rely on the concept of the "survival of the fittest." This involves creating a population of potential solutions and evolving them over generations to find the best one. Therefore, when you use Genetic Algorithms, the best solutions are kept, and the worst solutions are eliminated based on testing the various possible solutions. Genetic Algorithms have gained popularity in use in stock market forecasting applications.

Machine Learning

Machine learning encompasses techniques in which algorithms learn patterns from data and make predictions or decisions. It includes supervised learning where models are trained on labeled data, unsupervised learning where patterns are inferred from unlabeled data, and reinforcement learning where they learn to make decisions based on rewards or punishments.

Neural Networks

A Neural Network is a technology inspired by studies of the brain and nervous system. A network accepts several inputs, performs a series of operations on these inputs, and produces one or more outputs. The network accomplishes this by recognizing patterns in the data sets and using these to adjust internal weights and parameters. A structured network can classify new patterns or make forecasts and predictions based on the evolving patterns in the data.

The Neural Network, such as the General Regression Neural Network (GRNN), provides the capability to predict future activities based on past activities by making parameter adjustments reflecting a constantly changing data environment. The GRNN also provides statistics similar to those produced in a Multiple Regression Analysis, which can make the Neural Network explainable to others from a statistical perspective. Other popular processing approaches or architectures can be used to develop a Neural Network for a decision model.

A Neural Network provides the capability to incorporate up-to- date data from a variety of sources in the predictions, thus reflecting changes in operating characteristics of the business environment. This technology can be integrated into various system applications and easily operated by the user to provide a new type of analysis capability not currently used in the business environment.

Text Mining and Natural Language Processing (NLP)

Focused on processing and understanding human language, these are used for sentiment analysis, topic modeling, language generation, and more. Natural Language Processing attempts to capture information from text form and categorize this information for structured use in models. Many methods and approaches exist with these methods. Of personal note, these techniques are being used for big data analytics tools to utilize millions of patient records captured from health system electronic records. In many cases, the need exists to employ big data analytics tools to harvest critical information from these captured historical records from electronic health systems and augmented results with clinical research trials.

Evaluating Systems Performance with Statistical Analysis

Data analysis typically represents a great starting point in solving a problem. What does the data say is happening? Is this consistent with what people are saying is the problem? Are there inconsistencies between what people think and what the data shows? These basic statistical evaluations in an operating environment can confirm or deny what people think or can lead to a combination of truths that need to be further evaluated to get to the problem solution. It is worth your time to investigate your operating environment using statistical analysis. Data analysis is a critical component of quantitative decision making.

One caveat and consideration, however, is having a good understanding of the quality of the data that is used. For example, in a distribution center, is product being electronically scanned as part of being picked, packed and shipped? If you are looking to solve an inventory shrinkage problem, and only 98% of the electronically received data is actually being shipped to customers (accounting for inventory levels), then you may have a people problem with the management of the inventory in the warehouse. The data and preliminary data analysis can point you to actual issues in your organization, but you also need to talk with the individuals responsible for the data and information to truly understand what may be going on with issues in the information.

References

1. What is Big Data Analytics? Full Guide and Examples, Link: What is Big Data Analytics? Full Guide and Examples. Amplitude
2. Dovetail Editorial Team, Reviewed by Miroslav Damyanov, What is correlation analysis?, Last updated 11 May 2023, Link: What Is Correlation Analysis: Comprehensive Guide (dovetail.com)

Identifying Operational Improvement Methods

Abstract

Many methods such as Lean, Six Sigma, Kaizen and others assist you with identifying areas for improvement in your organization. To follow up with the identification of issues, you should employ optimization and operational efficiency methods to provide a robust framework for change and improvement. This chapter provides a high-level overview of some concrete operational methods that can be used to develop quantitative solutions. The methods, data used and decisions that are made with these methods are also presented. This provides a basis of methods that you can investigate further to improve your operations. These methods can serve as a foundation in quantitative decision making by integrating the data with robust methods and decision-making techniques to effectively model and improve your operating environment.

Identifying Improvement Areas

Data analysis is useful to identify specific areas where performance is not meeting targets and goals. However, data analysis alone may not lead you to methods and processes that sustain operational improvements. You should talk with the individuals that are performing the tasks to understand their issues and gain their input to develop meaningful area for improvement. Some areas identified may be obvious based on data. While data may show obvious improvements, some critical areas that may not be evident by data alone requires you to investigate processes to better understand the issues and opportunities for improvement.

Keeping in mind the overall purpose of these efforts, you want to identify operational improvements that support corporate objectives. Resources are almost always limited, so you need to focus your attention on those areas that support the organization strategically.

D. Cassone, *Operational Leadership Using Quantitative Decision Making*,
Synthesis Lectures on Engineering, Science, and Technology,
https://doi.org/10.1007/978-3-031-67565-2_5

Have an operational improvement mindset. Look at areas such as your budget performance, key performance indicators and customer service feedback. Spend time looking at your operations for better solutions. Also spend time talking to the individuals directly responsible for operations. Those individuals closest to actual operations may have the best ideas on how to improve performance. Be sure to give them credit in identified solutions, and they will be willing to share additional thoughts on other improvement areas.

Improvement focus areas should support the strategic objectives of the organization. Although other areas may be important, you will positively impact your organization if you can demonstrate your contribution to an organization based on how your improvements support the strategic direction of the organization.

Below are some well recognized and commonly used methods for identifying process improvements. There are many other methods that are not listed below. Importantly, as will be discussed later, is how do you model and implement those critical improvements based on quantifiably identifying those areas that most contribute to the overall objectives of your organization. It's important to spend your time and resources wisely [1].

- **Lean**—Techniques to view and eliminate waste.
- **Six Sigma**—Approaches to reduce the variability in processes.
- **Kaizen**—Focused activities on continuous improvement.
- **Total Quality Management (TQM)**—Product quality improvements with a heavy focus on your end-customer.
- **PDCA Cycle**—The Deming cycle, Plan, Do, Check, Act, for a stream of improvement initiatives.
- **BPM**—Business process management with the use of current state and future state business process mapping.

Your organization may have institutionalized one or more of these approaches with the end goal of improving operations. Great! Use those accepted methods for identifying improvements. Don't, however, focus on the method at the expense of the overall improvement objectives. As was noted in a previous chapter, you should be able to measure your opportunities and efforts with decision criteria and metrics that support corporate objectives and the strategic use of your limited resources. Be strategic in your use of resources.

Developing Incremental and Innovative Improvements

Once you are aware of or you have identified the problem, use quantitative methods to solve the problem. Many times, to truly solve the problem, simple fixes are not enough. You may need to employ more sophisticated and analytical techniques to model and generate solutions that go beyond the surface. Quantitative methods support quantitative decision making. Understanding the overall operating environment will guide you in applying the right analyses and techniques to solve the problem.

Optimization and Operational Efficiency Methods

Operational improvement may require more than process changes. More sophisticated methodologies may need to be employed to optimize operations at a larger scale. Examples may include adopting a new inventory replenishment strategy or implementing a route optimization program to minimize delivery travel time or fleet operational costs. The improvements identified by Lean, Six Sigma and other improvement identifications should be reviewed to see if there are quantitative tools and methods that are required to support those efforts at a larger scale. The main aim of the use of tools is to improve the productivity of the organization by optimum utilization of organizations resources: men, materials, and machines [2–4].

Industrial engineers are concerned with the effective utilization of organizational resources to maximize the productivity of an organization. Their training provides a broad base of course work that is designed to analyze and improve efficiency across an organization and improve an organization's operations. Tools and techniques taught in Industrial Engineering curriculums provide an excellent background that can be used in operational problem solving. These concepts provide a fundamental framework and thought process in gathering information, assessing information, and implementing solutions to support the strategic direction of the organization.

This chapter highlights tools that can be used for organizational improvements and real-world problem solving. Under each topic, sample methods are identified, data used in the application of the method and decisions made by the decision maker with the tools are presented. These topic areas are not comprehensive and are not addressed in detail in this book. Each of these topics are entire fields of study and with numerous resources readily available to better understand each topic. All of these methods help form a toolkit that can be used for quantitative decision making.

Facility Layout and Design

Facility layout and design involves designing industrial facilities for the efficient use and movement of material within the facility. Facility layout looks at the functions that are performed within a facility and locates machinery and materials in a way to maximize the productivity of an operation. Components of facility layout and design include the placement and movement of aspects of the industrial facility including equipment, materials, manpower, docks and staging areas.

Efficient facility layout and design reduces and eliminates bottlenecks within a facility. Entire courses are taught regarding different principles and techniques used in facility layout.

Sample Methods/Approaches

Methods and approaches include static approaches and dynamic approaches [5]. Dynamic approaches may include concepts such as a flexible layout, cyclical layout, and robust layout. Activity relationship diagrams are also used to identify requirements for the nearness of various operational functions. You most likely perform facility layout with a block layout phase along with a detailed layout phase to position man, materials, and machines.

Sample Data Supporting the Methods

Understanding the characteristics of your facility are important from both a physical characteristic and a material handling perspective. This includes understanding the number of facilities, number of floors, multi-dimensional view of the space, shape of the departments, area of the departments and dimensions of the departments. Expert opinion and computer aided packages are available to assist.

Sample Decisions Made Using the Methods

Numerous decisions are involved in facility layout exercises, and the implications or wrong decisions can be costly. Have you considered the capacity of the facility and its ability to support demand? Do you have an existing facility, or do you need to purchase or build a facility? What is your capital expenditure need for facility layout? Have equipment purchases been considered using robust industrial project evaluation methods? These types of decisions arise with facility layout decisions.

Engineering Economy and Industrial Project Evaluation

Typically, there is more than one alternative or potential solution to a problem. In the corporate environment, the short- and long-term cost and value of the alternative is important in making decisions. Decision makers must fully understand the ramifications of the costs associated with a dollar investment in an alternative. To do this, one must assess the alternative over a time horizon important to the decision makers. This involves assessing alternatives in terms of the time value of money, effects of taxation, depreciation, and inflation.

Sample Methods/Approaches

Methods used in engineering economy and industrial project evaluate the economic viability of industrial projects and the alternatives. These techniques assess the time value of money evaluating the future value, present value and annual value of alternatives and decisions on competing alternatives. Methods of evaluating competing alternatives by the rate of return, breakeven analysis, cost–benefit ratios are used. Additionally, risk is assessed, the impact of inflation and the evaluation of the minimum attractive rate of return across alternatives. Multiple criteria decision making may be used in these assessments.

Sample Data Supporting the Methods

Data used with these methods includes costs and revenues associated with equipment or projects, interest rates, inflation rates, alternative equipment configuration, strategic value of alternatives and risk.

Sample Decisions Made Using the Methods

Decisions made with these methods include make or buy decisions and selection of equipment alternatives. Capital expenditure decisions are made with these methods and many times, strategic considerations along with the financial evaluation are made using multiple criteria decision-making methods to evaluate alternatives in the context of broader strategic considerations.

Manufacturing Process Evaluation

A basic understanding of the types of manufacturing processes and systems is beneficial in operations management. Spending time on the production floor is one of the best methods to gain a working understanding of the types of work that is performed in an organization. It is also important to understand the relationships between product design and the manufacturability of a product when assessing and analyzing an environment. Good manufacturing processes and systems are key in the overall efficient functioning of an organization.

Sample Methods/Approaches

Strength of materials, manufacturing process theory, instrumentation, computer data acquisition, and data analysis concepts are used with manufacturing process evaluation. Laboratory testing of manufacturing processes and the engineering design of experiments for process variable measurements are used to develop efficient manufacturing processes. Six different types of manufacturing processes include job shop manufacturing, repetitive manufacturing, discrete manufacturing, batch process manufacturing, continuous process manufacturing, and 3D printing. Types of manufacturing styles include machining, joining, forming, and casting [6, 7].

Sample Data Supporting the Methods

Data used in manufacturing process evaluation includes product design, customer demand, financial resources, facility capacity, product characteristics, materials availability, personal and skill levels, industry and market considerations and data supporting the design, development, and manufacturing of specific products.

Sample Decisions Made Using the Methods

Decisions made around manufacturing processes are complex and must take into consideration many aspects of overall organizational operation. Should you produce your product on a manufacturing assembly line or is it specialized, and do you need more of a job shop approach to production? You need to consider the types of operations that are required to produce your product. Is it metal and does it need to be forged or is it plastic and can it be produced by a 3D printer? These considerations are impactful with regard to product quality and ultimately meeting customer demand.

Operations Research

Operations research is a scientific discipline that involves using mathematics and analytical principles to aid problem-solving and decision-making for organizations. Operations research is a subfield of applied mathematics that uses advanced mathematical principles or analytical methods to solve the problems that business leaders and managers often experience. Optimization and simulation methods facilitate modeling complex environments, improving operations and gaining insights into key aspects of organizational operations. As many operating environments are varied and complex, numerous optimization methods exist and can be used to best model your operating environment. Operations research encourages businesspeople to use advanced analytical tools to make more informed and effective decisions for organizations [8].

Sample Methods/Approaches

Methods include but are not limited to linear programming, network flows, graph optimization, integer programming, nonlinear programming, Markov chains, queuing theory, non-linear programming, normative theory of decision and games, dynamic programming, Markov decision systems, sensitivity analysis, and multiple-objective linear programming. This is a robust field of study, and these methods can be applied across an organization to model problems in various environments.

Sample Data Supporting the Methods

Data needed to develop optimization models tends to be comprehensive. Models numerically represent the objectives of the problem such as maximizing profit or minimizing cost, so you need data to represent the contribution to achieving objectives. Data is needed for the operating parameters of the problem being solved. The data needed for each type of optimization model will vary based on the use of the method and the actual operating environment. Products use resources and this typically will be modeled against constraints. Manufacturing has capacity constraints, and these constraints should be included in the model. Depending on the environment being modeled and the problem being solved, the variables that represent usage, capacity, and system requirements must be considered.

Sample Decisions Made Using the Methods

Decisions made with these types of models are vast and based on the purpose and construction of the model. If you are looking to maximize profit for a company, the model may be structured to determine the mix of products that you should produce to achieve maximum profit. If the model is used to minimize cost, you may use the model to determine the best transportation strategy for your organization. The decisions made with the models are as vast as the problems that are being modeled with the method.

Simulation

Simulation uses algorithms to evaluate the potential outcome of a situation, and it involves the creation of models to test various solutions before implementing them [8]. Simulation models many times provide more flexibility by allowing for the modeling and tailoring of the representation of complex environments. Basic concepts of computer simulation modeling can be applied to manufacturing, production, service, and other stochastic systems. Commercial simulation software is used to model the environment to build, analyze, verify and validate models. Simulation models are used in a variety of situations. You may need to model a production line to identify bottlenecks. You also may have a complex problem that does not fit into an optimization model that needs additional arrival and departure occurrences to be models. You may also need to test several scenarios that are represented in an operating scenario. Simulation models offer the necessary flexibility to represent real-world scenarios.

Sample Methods/Approaches

Types of simulation methods include stochastic simulation, discrete-event simulation, deterministic simulation, static simulation, dynamic simulation, and continuous simulation. Computer packages are typically used to formulate simulation models and execute them to determine outcomes.

Sample Data Supporting the Methods

Specific system components and flow are modeled using simulation. This may include the various steps in a manufacturing process, the movement of troops from one location to another, transportation vehicle movement, warehousing flow, customer arrival and queuing and many other model areas. Data representing each stage of a process represented with a simulation model is gathered and typically formulated into a statistical distribution so that random variables can be used to develop cases of the occurrences of the variety of conditions in the model. The data used to develop the representative statistical distributions can be vast, however, by representing this data in a statistical distribution, the data contained in the system can be minimal.

Sample Decisions Made Using the Methods

Decisions made with simulation models are as varied as the models themselves. Simulation models can be used to change production floor configuration prior to implementation. Simulation models can be used to represent pick, pack and ship processes in a warehouse and used to suggest new equipment or the use of additional resources to speed up the packing process. The military uses simulation models to make equipment and troop deployment decisions.

Project Management

Project management techniques provide a structured approach to proposing, planning and executing projects. You should have a good understanding of the fundamentals of project management to ensure project success. Tools and processes such as those listed below assist with robust project delivery and implementation.

- Project proposals.
- Project stakeholders.
- Scope management.
- Schedule management.
- Cost management.
- Change management.
- Communications management.
- Risk management.
- Resource management.
- Quality management.
- Procurement management.
- High performance teams.

The use of robust project management tools facilitates the successful completion and implementation of projects. Additionally, understanding appropriate use of approaches in project management can be used to best adapt your project to the nature of the specific project requirements [9, 10].

Sample Methods/Approaches

Complex projects with multiple interfaces may best be served by using waterfall and systems engineering project management approaches. Computer system design and execution may be best served by using Agile project management approaches. Hybrid approaches may be useful also, depending on the nature of the project. Understanding when you should use which approach is important in successful project management.

Sample Data Supporting the Methods

Project data typically includes a definition of the stakeholders, requirements, schedule data, cost data and resource utilization data. Additionally, detailed specifications may be developed along with an overarching project plan and a series of management plans to ensure that the project is completed on time, within budget and at the required scope. Plan to actual data will typically be tracked for in key components such as cost, schedule, and scope to assess the ongoing progress and performance of a project. Lessons learned and other key project performance data will be maintained to continually improve project implementation performance.

Sample Decisions Made Using the Methods

Decisions within project management will be around maintaining the project quality by meeting the scope, schedule, and cost project requirements. There may be a need to reduce scope based on schedule or cost constraints. You may need to expend extra resources to keep the project on schedule. Numerous decisions such as these will be made in the execution of a project.

Beyond the project execution decisions, an organization will decide whether or not it actually executes a project as part of its portfolio management and resource availability processes. Portfolio management processes will involve looking not only at the project scope, schedule, and cost, but also assess whether or not the project has strategic value and contributes to an organization's strategic direction and objectives. Resource availability may limit project initiation or require project rescoping. Multiple attribute decision making methods are helpful in making portfolio management decisions.

Production Planning and Inventory Control

Production planning and inventory control concepts involve understanding various methods to forecast production requirements and provide methodology for the control of inventory in an industrial facility. Numerous statistical and data analysis techniques are available to understand and manage production and inventory within a plant facility. Developing safety stock levels to minimize the times that product is not available within a facility can be dependent on both the production and the demand variability of the organization. The ability for a company to have product on hand or make an informed decision not to have product on hand, directly impacts the end-user level of satisfaction of the service of the company. Costs, production, and capacity considerations may impact an organization's ability to meet safety stock requirements. However, the ability or inability to supply product to customers directly impacts the bottom line of a company's operations in a number of ways. Using statistical analysis gives you insight on how to account for this variability.

You can translate statistical analysis of the variability of the process into safety stock levels maintained so that a supplier has a pre-determined amount of product on hand regardless of the demand and production variability. For example, you can set safety stock levels so that a product is on hand 95% of the time, based on the statistical analysis of the variability of the process. This would mean that you would run out of a given product only five times out of one hundred or 5% of the time, based on the average sales and the average amount of product maintained in inventory. Often, different products have different safety stock requirements. Management may state that the flagship products of a company should never be out of stock. You would set the safety stock levels for these products high in this case. However, when setting safety stock levels, you must also consider the amount of storage space required and cost for the product in inventory as well as the variability in production and sales.

Inventory replenishment methods are employed so that based on your safety stock requirements and the timing associated with receiving product once it is ordered are accounted for in inventory management. You must account for the variability in your customer demand patterns and the replenishment time that is necessary to have this product available on your production floor or in your warehouse. If you don't have product available, you may have to stop your production line or delay your shipments to customers, neither or which are beneficial situations. If you have too much product on hand, you will increase your inventory investment and occupy valuable warehouse space with overstocked product that may be needed for other products and components. Neither of these circumstances are beneficial to your corporate financial position.

Production planning is the process of deciding how a product or service will be manufactured before the manufacturing process begins. In other words, it is how you plan to manage your supply chain, raw materials, employees and the physical space where the manufacturing process takes place.

Production planning is very important for manufacturers as it affects other important aspects of their business such as supply chain management, production scheduling, material requirements planning, production lead time and capacity planning.

Sample Methods/Approaches

There are numerous methods that support the production planning and inventory control functions. These methods typically fall into the categories of job planning, batch planning, flow planning, process planning and mass production planning. Many times, Materials Requirements Planning software is used to facilitate the planning processes. Production planning operates in conjunction with a company's inventory management and replenishment strategy and whether the organization would like to make items to order or make items so that they are stocked and available to ship to meet customer demand. The planning process encompasses raw materials and finished goods, with an adequate supply of both needed to meet customer requirements. Production planning is an important part of overall supply chain management.

Sample Data Supporting the Methods

Data used to support these processes include comprehensive information about materials, labor, production methods, demand patterns, costs, and constraints. Production processes and inventory control are at the heart of a company's ability to deliver product to its customers and encompass large amounts of information required to support these processes. When you think, for example, how much material, parts, people, production processes, inventory planning, etc. goes into manufacturing a car, the amount of information required to meet customer demand may be overwhelming without a well-developed production planning system.

Sample Decisions Made Using the Methods

Many important decisions are made regarding production planning and inventory control. What is the company's fulfilment strategy? How much raw material and finished goods inventory should you have on hand to meet demand? Do you have the financial resources to maintain a large number of finished goods? Is technology changing and can finished goods become obsolete? Do you have sufficient production capacity? Do you have sufficient warehouse capacity? All of these decisions, although operational in nature, have a direct impact on overall corporate operations and the ability to achieve an organization's strategic goals.

Supply Chain Analysis and Logistics Engineering

Supply chain analysis and optimization is complex. Supply chain strategy and analysis comprise key components of overall organizational operations. These areas in general include those listed below:

- Procurement
- Inventory Management
- Forecasting and Planning
- Forward Logistics
- Reverse Logistics
- Order Management and the IT Infrastructure
- Vendor Management

- Logistics and transportation optimization
- Supply Chain Performance Management

A comprehensive view of the organizational supply chain is required to optimize its operation logistics system design, analysis, and management. Challenges arise in supply chain systems and quantitative models can be used to address these challenges.

Sample Methods/Approaches

The supply chain domain is large as it encompasses the management of its individual components and strategy overall. Procurement processes may be dictated by company policy. As discussed in the previous topic, inventory management, forecasting and planning have numerous methods to support its success. Forward logistics supports the optimization of the fulfilment and delivery of products to end-customers. Reverse logistics addresses the return of materials and the ability to repackage, repair and salvage materials from the return. An entire order management infrastructure is needed to receive orders, aggregate, and track them, send them to the warehouse or warehouse service providers and pick and pack orders and ready the orders for shipment. Optimizing the logistics and transportation functions can be very comprehensive. It can cost companies more than is necessary if done poorly and save companies significant money if done well. Optimization methods can be employed at various components of the supply chain to improve performance.

Sample Data Supporting the Methods

Data used to support supply chain operations is vast. Comprehensive supply chain management involves managing key aspects of procurement, inventory management, forecasting, forward logistics, reverse logistics, order management, warehouse management, transportation management, financing and vendor management. Supply chain scorecards and benchmarking can be used to support methods needed in a robust supply chain operation and measure performance. The Warehouse Education and Research Council (WERC) provides a robust list of logistics performance metrics that can be used for benchmarking to identify supply chain performance opportunities. [11]. They track key pieces of data that are required to support warehouse and supply chain operations. WERC groups key measures into sets which include metrics associated with customer, operational, financial, capacity/quality, and employee/safety, perfect order, and cash-to-cycle measurement.

Sample Decisions Made Using the Methods

The supply chain of an organization is its lifeblood. It supports the delivery of products to customers so that the organization can stay in business. Decisions around supply chain operations have strategic implications and should include the engagement of senior executives, from a strategic perspective. What are our on-time delivery requirements? What is our backorder policy? Should we outsource our warehouse operations? Do we need a new Enterprise Resource Planning System (ERP) to better support our organization and demand? The strategic requirements of an organization should then drive the operational performance requirements for those managing operations.

Statistical Quality Control

Statistical quality control concepts involve utilizing statistical techniques to monitor the quality of a production environment. This would also include quality engineering concepts and techniques, control charts, acceptance sampling, statistical process control quality data management and automatic inspection. These statistical techniques provide a means to keep the production environment producing items that are within the determined specifications. Poor product quality can be identified early when using statistical quality control methods and assist to avoid poor quality products getting to your customer.

Sample Methods/Approaches

Methods used to support statistical quality control broadly includes descriptive statistics, acceptance testing and quality control charts. Statistical methods such as the design of experiments may be used to support statistical quality control. Statistical quality control methods can also be extended to statistical process control.

Sample Data Supporting the Methods

Data used to support the methods will be based on the characteristics of the item being produced and monitored. If it is a food item, you may look at salt and seasoning content of the product. If it is a manufactured part, you may test the resulting dimensions of a milled or lathed part. The data will be assessed from a statistical analysis and the variability of product attributes will be charted and plotted to ensure that it is within acceptable tolerances.

Sample Decisions Made Using the Methods

Decisions made with statistical quality control techniques include determining acceptable tolerances, deciding whether or not to shut down a production line due to quality issues, sampling frequencies, etc. Statistical quality control monitors and controls the quality of components and end products and can be a critical aspect of the view of your company's quality in the marketplace and industry.

Systems Engineering and Analysis

Systems engineering is an interdisciplinary engineering management process used to develop an integrated, life-cycle balanced set of system solutions that satisfy customer needs. A systems engineering approach can be used to solve complex engineering systems problems. System engineering tools facilitate the development and life cycle management of complex systems.

Sample Methods/Approaches

Systems engineering process include analyzing customer's needs, requirements development, systems design, development, and integration, developing technical performance measures, system verification and managing cost, schedule, and risk in engineering tasks. Systems engineering project management plans include the development of an integrated master plan and integrated master schedule that contain the necessary detail to ensure that the development of the components and subsystems of a complex engineering system, i.e., space station, fighter jets, etc., are validated and verified for the delivery of a well operating system.

Sample Data Supporting the Methods

Data supporting systems engineering methods include information captured in traditional product and project development efforts, however, provide a focus on the interface requirements of complex systems. Requirements are developed for each of the system components along with their entrance and exit criteria requirements. Lifecycle costing data is necessary to capture the true costs of a system from conceptualization to disposal. Additionally, models, designs, documents, diagrams, drawings, and other developmental tools would be included as data that supports the systems engineering processes [12].

Sample Decisions Made Using the Methods

The following is a nice summary of system engineering decisions:

"Many systems engineering decisions are difficult because they include numerous stakeholders, multiple competing objectives, substantial uncertainty, and significant consequences. Decision situations (opportunities) are commonly encountered throughout a system's lifecycle. The decision management method most commonly employed by systems engineers is the trade study. Trade studies aim to define, measure, and assess shareholder and stakeholder value to facilitate the decision maker's search for an alternative that represents the best balance of competing objectives" [13].

Work Design and Ergonomics

Work design concepts are used to design and evaluate jobs in industrial working environments. There are many facets to work design. In general, the engineering design process involves defining the processes, analyzing them, determining potential solutions, evaluating alternatives, and selecting a solution. Techniques such as the use of Pareto Distributions can be used to identify the area with the greatest impact. Costs and return on investments can be used to analyze solution alternatives in the engineering design process. This can be applied in general developing solutions in many work environments.

Ergonomics is a field that works to optimize the interface between man and machine. This field addresses the human and machine aspect of a process and works to integrate the two into a comfortable working environment. Work physiology, biomechanics, design and organization of workstations, design of hand tools, climate, and sensory environment.

Ergonomic principles are surrounded around the concept of making the work environment efficient and comfortable to maximize work performance.

Sample Methods/Approaches

Tools and methods that may be employed in work design include statistical methods, value engineering, operations analysis between and within operations, time measurement systems (standard data systems, time and motion analysis, time measurement) and occurrence sampling. Ergonomics may incorporate methods around work physiology, biomechanics, design and organization of workstations, design of hand tools, climate, and sensory environment.

Sample Data Supporting the Methods

Broken down into a smaller unit, the individual functions within work are jobs performed either by individuals or equipment. A number of approaches are available to assess and aid in the job design process. The individual job assessments provide data that can be assessed and analyzed in determining the state of the given process. Data used to support work measurement includes standards, pre-defined time measurement techniques, work sampling, learning curves, line balancing and other data elements that help to define the work and environment being studied. Additionally, work physiology, biomechanics, climate, and sensory information may be used in work design.

Sample Decisions Made Using the Methods

Decisions surrounding ergonomics may involve the selection of one design over another design to best optimize the human–machine interface. Additionally, one may choose the height of objects to best meet a population demographic or the selection of a workspace design for an office remodel. You may also choose to automate additional processes due to difficult manual construction requirements or additional time operations take that may better be handled by a machine.

Decision-Making Methodologies

Numerous decision-making methodologies exist and can be integrated into quantitative decision-making solutions. The fields of multiple attribute decision making, multiple objective decision making, and group decision making are vast and contain numerous methods that can be applied in problem solving. These methods are important because they integrate the decision maker's perspective into the overall decision and improvement process to ensure that the end solution will meet the overall organizational objectives. Decision making methodologies along with statistical analysis and methods are critical in quantitative decision making and both can be applied to the methods discussed in this chapter. The following chapter will discuss decision making methods and a few specific methods in more detail.

Quantitative Methods Across Engineering Disciplines

The author is most familiar with Industrial Engineering methods, based on her education and application in the workforce and consulting engagements. All engineering disciplines and many other disciplines learn, apply, and implement quantitative methods in their job functions. This may include principles of machine design, statics and dynamics principles in road construction, fluid dynamics in chemical engineering and the design of electrical circuits and controls in electrical engineering. These areas will not be covered in this text, however, all these types of quantitative methods integrate into quantitative decision modeling. These sound quantitative principles form a robust framework for quantitative decision modeling and decision making.

References

1. PMO Team Clickup.com, What Are Process Improvement Methodologies? 6 Types to Know, last updated September 29, 2023 Link: What Are Process Improvement Methodologies? 6 Types to Know (clickup.com)
2. Raaz, Md. Noor Nabi, Tools & Techniques of Industrial Engineering Used in Apparel Industry, Last Updated on 06/09/2021, Link: Tools & Techniques of Industrial Engineering Used in Apparel Industry (textilelearner.net)
3. Editor(s):Gavriel Salvendy Ph.D., Handbook of Industrial Engineering: Technology and Operations Management First published:11 May 200, John Wiley & Sons, Inc., Hoboken, New Jersey
4. Bidanda, Bopaya, Maynard's Industrial and Systems Engineering Handbook, 6th edition, McGraw Hill, New York, New York, October 12, 2022
5. Pablo Pérez-Gosende, Josefa Mula & Manuel Díaz-Madroñero (2021) Facility layout planning. An extended literature review, International Journal of Production Research, 59:12, 3777-3816, https://doi.org/10.1080/00207543.2021.1897176
6. Lauri, Karl H, Production Planning Methods, Published February 4, 2020, Production Planning Methods - Blog for Manufacturers and Distributors (mrpeasy.com)
7. Indeed Editorial Team, 6 Types of Manufacturing Processes, Updated March 10, 2023 6 Types of Manufacturing Processes. Indeed.com
8. Indeed Editorial Team, What Is Operations Research? (Definition and Examples), Updated January 3, 2023 Link: What Is Operations Research? (Definition and Examples). Indeed.com
9. Kerzner, Harold, Project Management: A Systems Approach to Planning, Scheduling, and Controlling, 13th Edition, John Wiley and Sons, 2022
10. Verzuh, Eric, The Fast Forward MBA in Project Management, Sixth Edition, John Wiley and Sons, 2021
11. Metrics - Warehousing Education and Research Council (werc.org)
12. Wheatcraft, Louis S., Ryan, Michael J., Svensson, Carl, Integrated Data as the Foundation of Systems Engineering, INCOS International Symposium, First Published 30 August 2017, https://doi.org/10.1002/j.2334-5837.2017.00438.x

13. Madachy, Ray, Roedler, Garry, Parnell, Greg, Jackson, Scott, Decision Management, Guide to the Systems Engineering Body of Knowledge (SeBOK), Downloaded 4/1/2024, https://sebokwiki.org/wiki/Decision_Management, 4/1/2024

Applying Decision Methodologies

Abstract

Organizational decision makers must be included in the quantitative decision-making process. If decision makers are not engaged in operational improvement efforts, they can impact progress in accomplishing organizational goals. This chapter discusses some general aspects of multiple criteria decision making and group decision making that can be used to integrate decision makers into the overall decision process. Additionally, recommendations for easy to use and easy to understand methods that can be employed are made. The field of decision making, and decision science is vast, so this chapter is only intended as a brief introduction and to include practical methods that are useful in the context of quantitative decision making.

Introduction

To develop robust solutions, corporate objectives and decision-makers need to be integrated into the decision-making process. Methods exist so that corporate objectives and executive decision makers have an ongoing voice in the solutions that are developed. This is done by incorporating decision-making techniques into the solution model.

This section provides an overview of decision-making methods that are practical and useful in the organizational environment. The methods discussed in this section have been used to solve real-world problems in industry, the government, and the military. Decision making methods can be simple or complex. The methods discussed in this chapter are not complex, because the author has found through numerous military, government and corporate projects that the decision maker and their understanding of the method was critical to deliver a solution that would be used by the decision maker. If decision-makers did not understand the method being used, they were reluctant to use it. This drove a simpler

D. Cassone, *Operational Leadership Using Quantitative Decision Making*,
Synthesis Lectures on Engineering, Science, and Technology,
https://doi.org/10.1007/978-3-031-67565-2_6

application of decision-making methods to ensure that the objectives of the organization and the decision-makers preferences were included in the process. Methods that were more complex were more difficult when justifying the results and explaining the choices delivered by the quantitative decision-making model to senior levels of the organization. For this reason, you will see a more simplified approach to applying decision-making methods to develop solutions.

The application of simpler decision-making methods comes after the study of complex decision-making methods. A number of years ago, Dr. Ching-Lai Hwang published books on Multiple Attribute Decision Making, Multiple Objective Decision Making and Group Decision Making [1–3] in both crisp and fuzzy environments. Additionally, Dr. Hwang developed the Technique for Order Preference by Similarity to the Ideal Solution (TOPSIS) [3] method which is a widely used multiple attribute decision making method and is discussed in this section. Understanding these methods provided a framework to select and use the most appropriate decision-making methods to solve specific problems. It is a powerful combination of theory and application when you have the knowledge of the methods, understand the operating environment, can select appropriate tools to solve problems and then encapsulate decision making methods in a delivered solution.

Practical Multiple Attribute Decision Making Methods

Multiple Criteria Decision Making

Multiple criteria (also known as multi-criteria) decision-making (MCDM) is a well-known branch of decision making. Multiple criteria decision making is typically divided into two classes of problems, the first is multiple objective decision making (MODM) and the second is Multiple Attribute Decision Making (MADM). MODM models are typically more sophisticated mathematical models involving linear programming fundamentals but are methods that support multiple conflicting objective functions that can be used to trade-off the decisions space supported by the model. The objective functions or objectives of the model may or may not have equal importance and MODM methods can handle the differences in importance of the objectives and the continuous design nature of the linear programming model. The nature of MODM problems differ somewhat from MADM problems in that their form and structure in that MODM problems involve designing the best alternative given a set of conflicting objectives. A MODM problem involves developing a model that has conflicting objectives subject to resource constraints and will allow for an infinite number of design possibilities.

Multiple Attribute Decision Making (MADM) refers to making preference decisions over the available alternatives that are characterized by multiple, usually conflicting attributes. MADM problems in their base form have a discrete number of alternatives that have multiple characteristics or attributes. The units of the multiple attributes are typically incommensurable units and each of the attributes associated with the alternatives may have a different importance in the decision or weight of importance in the decision [4].

There have been many developments throughout the years in the field of Multiple Criteria Decision Making. The author prefers to reference some of the base methods that were presented by Dr. Hwang in his early books as a more comprehensive list may not be necessary in an applied environment. These methods are used to support operational leadership using quantitative decision-making and to support the overarching decision-making process. Figures 1 and 2 show a limited taxonomy of Multiple Objective Decision-Making methods and Multiple Attribute Decision making methods, respectively.

Multiple criteria decision making is a broad and deep field of study and is not intended to be addressed in detail in this book. The intent is to make the reader aware of these methods and techniques and how these methods can be useful in quantitative decision making. The decision maker is key in your operating environment and using structured methods that can be integrated into the decision process in an operating environment and support meeting corporate objectives.

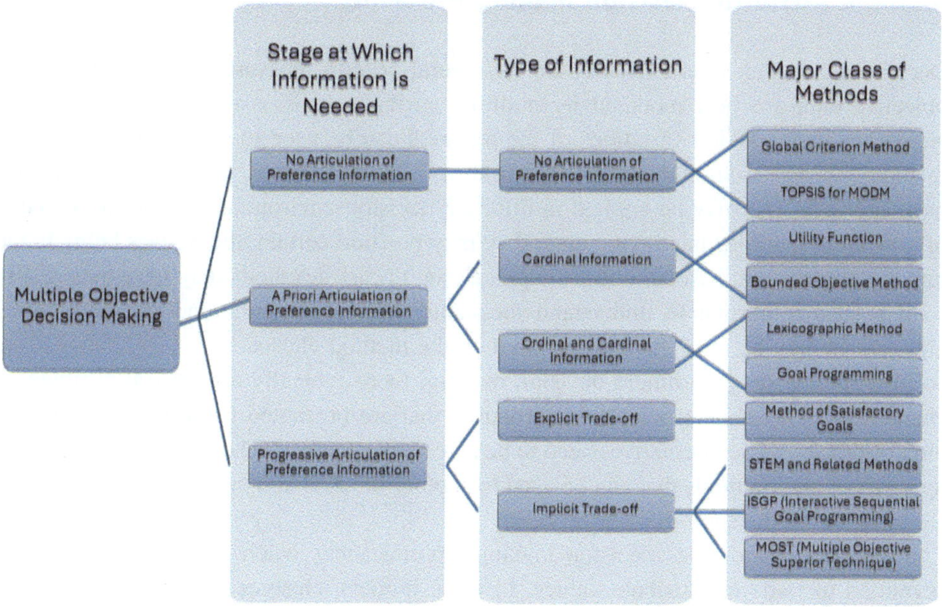

Fig. 1 A taxonomy of MODM methods (Adapted from Hwang and Yoon [3]

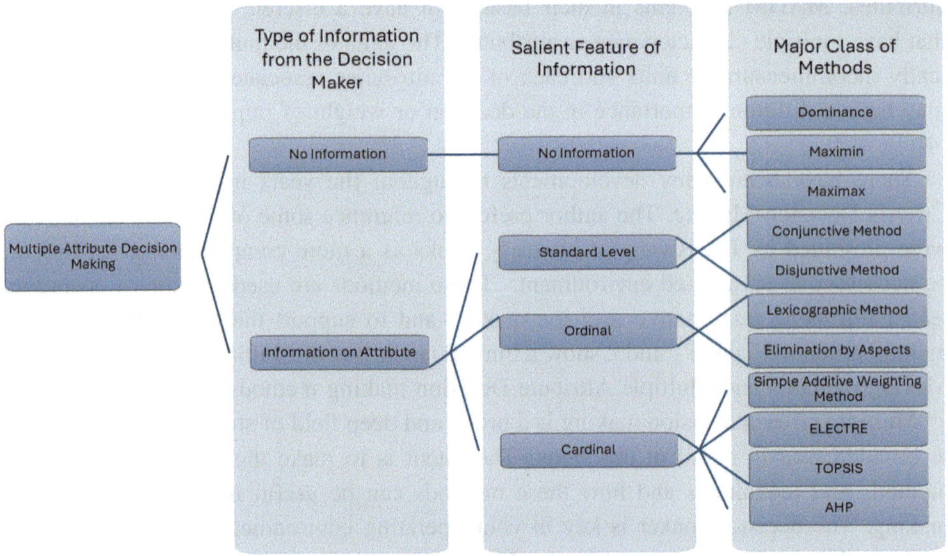

Fig. 2 A taxonomy of MADM methods (Adapted from Hwang and Yoon [3]

Multiple Attribute Decision Making Problems

Decisions that rank alternatives based on several criteria measured with subjective and objective data are best modeled by multiple attribute decision-making methodologies. Numerous methods and variations of the methods can be used to rank alternatives. Each method and its variation process the data in a different manner. Each of these methods have strengths and may be applied in different decision environments. Due to the nature of how these methods handle the data, they may position certain alternatives being higher ranking alternatives, based on how the processing occurs. Methods may have biases, also, so it is also important to understand the nature of the computations and how certain alternatives may be more favorable, based on the method chosen.

In corporate or government decision making, there typically are several attributes or characteristics associated with the alternatives that are presented for a decision. In many circumstances, these alternatives need to be screened or ranked. There are several different approaches and methods that can be used to capture the decision environment and provide structure to the decision-making process.

This section addresses decision making circumstances where the decision maker is permitted to trade-off attribute values. In these models, changes in one attribute or the weighting associated with that attribute can be offset with changes in any other attribute. With models such as these, a single number is usually assigned to each multidimensional characterization representing an alternative.

Additionally, the data used in the decision is very important in the overall decision process. Subjective data is typically forward-looking when compared to objective data, which is historical or looking back. If some objective or historical data are readily available, this can be analyzed with statistics and used, along with the subjective data that is generated with expert opinion and used as input to the model. Numerous methods exist and can be applied to multiple criteria decision-making models, such as Simple Additive Weights (SAW) or the Technique for Order Preference by Similarity to Ideal Solution (TOPSIS). These two methods are easiest to apply and easiest to understand and provide the best solution.

Examples of multiple criteria decision-making models may include the following:

- Selection of executives for promotion or retirement based on performance, evaluation criteria, and need of the organization.
- Ranking critical items for the military based on their contributions to operation plans, readiness, sustainability, and availability.
- Selecting products to keep or delete in a product line based on sales volume, future potential sales, strategic importance, and their impact on operations.

If the time frame for the decision to be made is longer, you can use more sophisticated models to solve the problem. Based on the business conditions, the availability of data, the condition of data, the number of objectives of key decision makers, and the organization's goals, you can combine methods to develop the model [5, 6].

Other features of the multiple attribute decision making problem include an attribute hierarchy, conflict among criteria, a hybrid nature, uncertainty, a large-scale problem, an ideal-solution, a non-dominated solution, a satisfying solution and preferred solution.

The Decision Matrix

The decision matrix provides a structure to solve multiple attribute decision making problems and capture the key aspects of the decision being considered. A MADM problem may be described using a decision matrix. This matrix can be used with various MADM methods and forms the framework for capturing the key elements being considered in the decision.

For a MADM problem, suppose there are m alternatives to be assessed based on n attributes. A decision matrix is a $m \times n$ matrix with each element a_{ij} being the j-th attribute value of the i-th alternative. Decision criteria or attributes are represented as C_j. Importance weighting associated with the decision criteria is represented at w_j and the alternatives under consideration are represented by A_i. Table 1 is a representation of the decision matrix and a brief definition of the elements of the matrix.

- *Criteria (Attribute)*: A criterion or attribute is a measure of effectiveness. It is the basis for evaluation. In multiple attribute decision making problems they can be viewed

Table 1 The decision matrix

Weights	W_1	W_2	...	W_n
Alternatives/criteria	C_1	C_2	...	C_n
A_1	a_{11}	a_{12}	...	a_{1n}
A_2	a_{21}	a_{22}	...	a_{2n}
...	...	...	...	...
A_m	a_{m1}	a_{m2}		a_{mn}

as performance parameters, components, factors, characteristics, and properties. An attribute should provide a means of evaluating the level of achieving an objective. Each alternative can be characterized by a number of attributes, based on those elements that are important in the decision process.

- *Weights*: Not all attributes are equally important when making a decision. Minimizing cost, for example, may be far more important than maximizing comfort. By assigning weights to the decision criteria or attributes, the decision maker or team can represent their value of importance in the ultimate decision. As a note, not all decision methodologies support criteria weighting, therefore it is necessary to select a method that is representative of the types of considerations that the decision maker or makers wish to consider.
- *Alternatives*: Alternatives are those potential choices that a decision maker can select based on his or her evaluation of the key aspects of the decision. You, for example, may be deciding to purchase a car and use a decision matrix for your evaluation process. Your alternatives are the different types of cars that you are considering, and the decision criteria or attributes are those elements that are important to you in your decision-making process.
- *Attribute Value for Alternatives*: These values are the representation of the performance level for an alternative as it is associated with each attribute. For a given alternative, there is a scoring of each attribute to represent the value and consideration associated with the alternative. These evaluations are the data that are used in the evaluation process.

The decision matrix is a fundamental component of multiple attribute decision making problems and is used as a basis for applying numerous different types of decision-making methods to solve multiple attribute decision making problems. The development of the attributes, alternatives, importance weighting and data used in the decision matrix is important to developing a well-structured decision problem.

Data Used in the Decision Matrix

You can use either objective or subjective data to represent metrics used in the decision process. Objective data usually can be described as data that can be quantified by some measure of known commonality. This may be data such as the number of items produced, number of trucks in a location, population of a city, and so on. This data is usually available in some form in company databases and information systems. Typically, statistics such as averages and trends are generated based on a record of this objective data over some period in time. Objective or quantitative data represents a history of activities of a company that has been operating during a given time period.

Qualitative or subjective data can be easily used in a number of different situations. Surveys are good examples of subjective data used to represent a rating of a product or service. Use scales from 1 to 5 or 1 to 10 to represent high, medium, and low assessments for given metrics. Use assessments such as red, yellow, and green in other situations in which individuals (such as military personnel) might find more meaning in rating conditions. Numerical values with their verbal description provide the type of information that can be captured and utilized in a decision model when other information is not available.

Subjective data is data based on someone's opinion or best guess of a condition or a future event. Subjective data is more qualitative in nature in that it defines a situation or condition without specific data points. Subjective data can be generated by individuals within or outside of a company or experts within a given field of operation. Subjective data and expert opinions are typically forward looking in nature trying to predict what will happen in the future. Individuals make assessments based on what has happened in the past and what may happen in the future.

Objective data, especially in the form of statistics, however, is based on historical data, thus projecting the future, what has happened in the past, which assumes that the future will behave much like the past. The entire business environment may have changed; thus, what has happened in the past may be a poor representation of the future; thus, a new source of data is required.

In developing decision criteria or attributes that are used in multiple attribute decision making problems, you should consider a number of parameters in the development process to ensure a set of well-structured, well-represented goals and decision criteria. These development parameters are as follows:

- Goals and decision criteria must represent actual and important considerations in making decisions.
- Decision criteria must differentiate one project from another in terms of higher or lower priority.
- Decision criteria must be independent, not overlapping in content or intent, to avoid accounting for the same thought or idea more than once.
- Decision criteria must be defined as clearly as possible to ensure that the decision criteria in the evaluation process are viewed in the same context.

- Measures and scales developed for the decision criteria must be meaningful in the evaluation process and the data to perform the evaluation easily accessible.
- Constraints that represent types of mixes, qualifiers, and conditions that would be applied to a prioritized list of items must be identified and differentiated from the evaluation decision criteria.

These are important considerations in the construction of a decision matrix for decision making. The quality and meaningfulness of the data used in the decision process is paramount in the quality of the decisions that are made by the model [5, 6].

Simple Additive Weighting (SAW)

The Simple Additive Weighting method (SAW) is probably the best-known and widely used method of multiple attribute decision making. To each of the attributes in SAW, the decision maker assigns importance weights, which become the coefficients of the variables. These weighted coefficients need to be normalized. To reflect the decision maker's marginal worth assessments within attributes, the decision maker also makes a numerical scaling of intra-attribute values. The decision maker can then obtain a total score for each alternative simply by multiplying the scale rating for each attribute value by the importance weight assigned to the attribute, and then summing these products over all attributes. After the total scores are computed for each alternative, the alternative with the highest score (the highest weighted average) is the one prescribed to the decision maker.

There are some precautions that are important when developing decision models.

1. Scaling of criteria value can greatly influence the impact of single criteria and thus the ranking. To avoid this problem, all values within a criterion are normalized or a similar scale can be developed for each of the criterion.
2. Independence: Care should be taken so that all the criteria are independent, thus avoiding overweighing the effects of a single criterion.

Technique for Order Preference by Similarity to Ideal Solution (TOPSIS)

The Technique for Order Preference by Similarity to Ideal Solution (TOPSIS) is a multiple attribute decision-making problem with a number of alternatives evaluated by a number of attributes. The TOPSIS methodology is viewed as a geometric system with m points in the n-dimensional space. TOPSIS is based on the concept that the chosen alternative should have the shortest distance to the best possible attribute assessment values (positive-ideal solution), and have the longest distance from the worst possible assessment values (negative-ideal solution) for each of the alternatives.

An ideal solution is defined as a collection of ideal levels for the attribute levels. An ideal solution, however, is ideal and usually not attainable. The TOPSIS methodology uses concepts to come closest to the best possible answers and furthest from the worst possible answers. The positive-ideal and negative-ideal solutions are used in this methodology to find the best feasible alternative given the options and evaluations available. This method has a strong mathematical basis and has proven to be robust and well accepted for ranking applications in industry, the government and the military.

Sensitivity Analysis

The models and results provide a means to identify the key drivers associated with the assessment of technology development alternatives. You can perform sensitivity analysis on the current ranked list of technology alternatives to determine how much an alternative evaluation criterion must change to move an alternative up or down in the ranked list of projects. The importance weighting of the evaluation criteria and the evaluation scores for the alternatives drive the alternative scoring in terms of meeting customer and organizational goals. Sensitivity analysis is useful in identifying the key drivers of the overall value of an alternative. The specific sub-criteria evaluations for the alternatives provide the supporting detail to show why certain alternatives received the scores they did and can be used to pinpoint potential areas for improvement and additional trades.

The methodology used to perform sensitivity analysis in MADM problems consists of testing each criteria for each alternative to determine what the value would need to be to move the alternative up or down in rank. To determine the values of the criteria for the alternatives required to move the alternative up in rank, the criteria is incremented one increment closer to the best value for that criteria. To determine the values of the criteria for the alternatives required to move the alternative down in rank, the criteria is incremented one increment closer to the worst value for that criterion. The model is then run, and the results are tested against the original ranking to determine if the change in this criteria value has moved the alternative up (or down) in the rank of the alternatives. The incrementing process continues until a value is determined that can move the criteria up (or down) in rank and is done for each of the criteria for all alternatives. Also, the incrementing process is done on only one criterion value at a time, thus testing each criterion value independent of any other changes made to the other criteria values [5, 6].

Practical Group Decision Making Methods

Introduction

Many times it is necessary to get the consensus of a group of executives and extract their ideas and expert opinion. Group decision making has been used to determine the objectives that most satisfy the goals of the organization. These objectives can and do change over time, and they change when the makeup of the group changes. Each can be used to gain a group consensus of goals and objectives. You can apply different methods for group decision making, based on the expertise, dominance, and political nature of the various decision makers involved in the process.

It is also useful to use group decision making when only subjective data is available about the future to support and develop a model. Because it is hard to look into the future, sometimes the one resource to get a forecast of future events is to poll experts in the field, i.e., get expert opinion. Group decision making is useful to use when the problem is complex and difficult to define. The "gut feel" of experts can provide good overall information to support a model when no other data is available. The time frame for the decision can dictate the type of model and the type of data that can be gathered for use in the model. Well-structured data and information gathered in group decision making can be used in analytical models. This expertise captured can be combined with objective data to provide a well-represented set of information to solve problems.

Types of Decisions Made by an Organization

Each level and function of an organization is involved in decision making, in some general capacity. Some of these decisions may be made by individuals on a daily, ongoing basis, while others are made by a group to come to a decision on a key matter facing the organization. Routine or repetitive decisions are made, which may or may not require group decision making or group consensus. Decisions are made where there is no easy solution and may require executive level review and agreement.

Strategic decisions are typically made at executive levels of the organization and dictate the direction and activities that move an organization into a future position. Some decisions may be more of a tactical nature or impact policies within an organization. Organizational structure and alignment decisions may be made that could include reorganization or reduction of staff. All of these types of decisions may have major or minor impact and will require varying levels of engagement from individuals in the hierarchical positions of the organization.

Since organizations make so many decisions on a day-to-day basis, it is good to consider how those decisions are made. Understanding and being aware of the types of decisions made by an organization is helpful in looking for opportunities to apply structured group decision making techniques. It is important to not only know and understand your decision makers, but to also look at these decision makers to determine what types of decision-making methods may be most beneficial in their decision-making process. The next section of the section will begin to describe the classification of group decision making methods and general areas where these methods can be applied.

Classification of Group Decision Making Methods

Group decision making includes diverse and interconnected fields in the areas of preference analysis, utility theory, social choice theory, committee decision theory, theory of voting, general game theory, expert evaluation analysis and many others. The goal with group decision making is usually understood to be the reduction of different individual preferences among objects in a given set to a single collective preference, or group preference. The members of the group can use many different approaches or a combination of approaches to arrive at the single collective group preferences. These approaches can be divided into three general fields which are social choice theory, expert judgement/group participation and game theory.

Social choice theory involves three broad classifications of decision making which is comprised of methods that handle voting, social choice function and social welfare function. Voting is a group decision making method in a democratic society which is an expression of the will of the majority. It is a multiple criteria decision-making process whenever a voter casts a vote to select a candidate or policy. The candidates' qualifications may be judged by multiple criteria such as trustworthiness, honesty, capabilities, political stance and positions on specific issues. These criteria are summarized in a voter's mind, into a value function that is represented by their selection for a candidate. With single issue voting, the voter considers the single issue as most important over the other issues. The voter rates the candidates in their choices in regard to their value function.

Social choice functions can be viewed as aggregation procedures based on preferential voting systems representing social choice. A social choice function is a mapping which presents a subset of the potential feasible set of choices into a preferential mapping of individual's preference profile. Condorcet's principle, discussed in this section, provides a great example of the aggregation of preferential ranking of individuals. The Borda function provides this mapping as well.

Social welfare function involves defining fair methods for merging the individual preferences into a societal choice. The social welfare function is comprised of a rule which must map the profile of individual preference orderings into one of the possible preference orderings for society itself. Social welfare functions are derived based on several conditions and can be studied from a theoretical perspective.

Expert judgement and/or group participation is comprised of methods that include brainstorming, brainwriting, nominal group techniques, surveys, the Delphi method, conferences, successive proportional additive numeration (SPAN) and other modeling, simulation, and cognitive mapping techniques. The method selected is determined by circumstances that involve data availability and/or the operating environment. Sometimes the method that the decision maker feels most comfortable with is selected. It is easier to explain to him and for him to be comfortable with its use.

Game theory uses different forms of the game including the normal form of the game, characteristic function form, cooperative and noncooperative games, zero-sum, constant-sum, and general sum games being grounded in the concepts of the Nash Equilibrium. Game theory is a mathematical technique used in analyzing conflict-of-interest situations. Game theory is concerned with individuals who are pursuing their own self-interest and personal values against other individuals who are pursuing their own self-interest and personal values. Game theory can be mathematically complex and challenging, but its underlying assumptions are simplistic. Large scale games may also be difficult to apply in real-world decision problems, however, investigating the concepts and principles in game theory can be interesting. This topic, however, will not be addressed in this section [5, 6].

Group Decision Making Methods

Condorcet's Principle

In analyzing voting from the Condorcet Principle perspective, there is only one rigorous way of knowing the wish of the majority in an election. This consists of basing this wish on the respective merits of all candidates compared one by one. The candidate should be elected if he beats every other candidate under simple majority when such a candidate exists. This principle is based in part on an argument involving probabilities of correct judgements and embodies the democratic precept of rule by majority will. The majority candidate constitutes a stable equilibrium in that it cannot be beaten by a challenger in a direct majority vote between the two. The concept behind the application of the Condorcet Principle is similar to a maximin function since it chooses those candidates whose worst showing against the others is as good as possible.

In determining the best rank or preferences, a problem sometimes occurs in which A is preferred to B, which is preferred to C, and where C is preferred to A. This occurs when we directly compare alternatives to one another without using criteria. In the above preference, each alternative is scored against all others. Condorcet suggests a method to break the cycle and to rank the alternatives. It determines the worst each candidate does against all other alternatives. Then rank each alternative as the best of the worst, so that it is a ranking by the best of the overall worst values.

The Borda Function

The approach that Borda suggests is that we rank each candidate or team 1 through *n*. Each participant ranks the group of candidates 1 through *n* and then adds their rank scores (team scores) across each participant. The lowest score is ranked first and so on to the highest score, which is ranked last. This approach is used in the coaches' poll of football rankings, where each coach ranks all teams 1 through *n* and then adds across each team and the one with the lowest score is ranked first.

The following example shows how this function works. Three different coaches have ranked seven different baseball teams. Tables 2, 3 and 4 show the rankings of each coach, respectively.

The aggregate rankings and results from the coaches' rankings are shown in Table 5.

The rankings by the coaches are summed across each team. For example, Coach 1 ranked the Mets 3, Coach 2 ranked the Mets 4 and Coach 3 ranked the Mets 2. The sum

Table 2 Coach 1's team ranking

Team	Coach 1
Royals	1
Cubs	2
Mets	3
Blue Jays	4
Twins	5
Cardinals	6
Yankees	7

Table 3 Coach 2's team rankings

Team	Coach 2
Yankees	1
Cubs	2
Royals	3
Mets	4
Blue Jays	5
Cardinals	6
Twins	7

Table 4 Coach 3's team rankings

Team	Coach 3
Blue Jays	1
Mets	2
Twins	3
Royals	4
Yankees	5
Cardinals	6
Cubs	7

across the coaches for the Mets is a total of 9. When we do this for all of the teams, we see in the example that the Mets ranked second overall with the list of teams, being beat out by the Royals. Obviously, this ranking can vary significantly by coaches over time.

You may also note that you can also do the ranking in reverse order where you give the highest ranked option the most points, i.e., where there are m candidates, you would assign marks of n-1, n-2, …, 1, 0 to the first ranked, second ranked,…, last ranked candidate for each individual then determine the Borda score for each candidate as the sum of the

Table 5 Aggregate results and rankings

Team	Coach 1	Coach 2	Coach 3	Sum	Rank
Royals	1	3	4	8	1
Mets	3	4	2	9	2
Blue Jays	4	5	1	10	3
Cubs	2	2	7	11	4
Yankees	7	1	5	13	5
Twins	5	7	3	15	6
Cardinals	6	6	6	18	7

individual marks for that candidate. In this case, the candidate with the highest Borda score would win.

Expert Opinion/Group Participation

Decision makers and problem solvers in government, business, industry, and education are confronted with a variety of problems. These problems are highly complex, often interdisciplinary, with social, economic, political, and emotional factors being considered alongside more quantitative data. When solving these problems, it is important to account and consider all of these factors, whether you initially feel hard, quantifiable data is available.

Group decision making, as stated before, is understood to be the reduction of many different individual preferences into a single choice either by conflict or compromise. High-quality decision making requires the decision maker to see through the problem and its complexities and apply decision-making methods best representative of the decision environment and of the key decision makers involved in the decision-making process. Providing a structure to support quality decision making is important.

The approaches in the following sections represent an important group that can be used to assist decision makers in the decision-making process. These methods support the extraction of ideas by idea stimulation and creative confrontation and issue clarification. Experts are then polled for their evaluation, classification, and selection of courses of action [5, 6].

Brainstorming

Brainstorming is a widely used method for idea generation. It is most effective when a set of rules is understood and followed. The first rule is to defer judgment, so ideas are generated freely, and the participants feed on one another's ideas and generate a larger number of possibilities. This is done to delay judgment, which stymies new thoughts. Judgments can be made later. Another reason is that the more ideas produced, the better the ideas; thus, with more ideas generated, there is a higher probability of finding the best solution. The judgment of ideas and the solution is completed after most ideas are generated. If it is done too soon, it stymies the flow of ideas.

Brainwriting

The process of brainwriting is similar to brainstorming, but ideas are written down rather than communicating them orally with the leader. All ideas are written down on a sheet of paper and put on the center of the table. Then the leader takes the paper with the ideas written on it and it is passed around and added to by each participant. After a period of time, say thirty minutes, the sheets are gathered for later evaluation. Using this procedure, all participants can share their ideas on an equal basis. A dominant personality will not be as likely to take over the process and push only their ideas. Reading others' ideas on paper, the process is continued until the improvement of ideas stops. The process is not suitable for large groups. Again, the rules of the process are reviewed with the participants, and they are the same as in brainstorming.

The Nominal Group Technique (NGT)

Nominal group technique (NGT) combines elements of brainwriting, brainstorming, and voting techniques. It creates a robust method that is widely used for idea-generation and problem-solving and for a wide variety of applications. The name "nominal" comes from considering a collection of individuals as a group "in name only" or "nominal," since there is no verbal exchange permitted. Generating ideas nominally in a group minimizes confirming influences. It ensures all participants can contribute equally in the group decision.

Surveys

Surveys for problem solving are used to quickly gather data that is not normally available and to get a better understanding of the environment. It also provides insight into the process details. Surveys are also useful in getting unavailable data quickly for problem analysis. Using the Internet for your survey forms is a good way to quickly get the results and classify and analyze these results. Since the process is quick and data is so easily gathered, surveys should be used more frequently to track the progress of process improvement efforts.

Delphi Method

The Delphi Method is a combination of brainwriting and survey methods. The object of this method is getting a reliable consensus of a group of experts. It attempts to do this through a series of detailed questionnaires along with expert opinion feedback. The Delphi method has several components: anonymous participation, controlled iterations, feedback, and statistical results. The number of iterations of the questioner is normally three to five. Additional iterations depend upon the degree of agreement and the direction of the respondents from their feedback information.

Summary

The goal of Group Decision Making is to get a group consensus. You can apply numerous methods in Group Decision Making, all of which are designed to assist in the extraction of ideas, structuring problems and arriving at a solution that the group involved will be satisfied with overall. The field is broad and there are many methods available to use to assist in group decision making. You can also apply different methods for Group Decision Making based on the expertise, dominance, and political nature of the various decision makers involved in the process. This section presents methods that have been tested and used by the author and found to be beneficial in real-world application [5, 6]. The goal of this chapter is to present you with methods that engage decision makers, incorporate they preferences into the decision process and provide structure for your organizational leadership decision-making processes.

References

1. Hwang, Ching-Lai, and Masud, Abu Syed Md. *Multiple Objective Decision Making—Methods and Applications*, Springer-Verlag, Berlin, Germany, 1979.
2. Hwang, Ching-Lai, and Ming-Jen Lin. *Group Decision Making Under Multiple Criteria*, Springer-Verlag, Berlin, Germany, 1987.
3. Hwang, Ching-Lai, and K. Yoon. *Multiple Attribute Decision Making Methods and Applications*, A State-of-the-Art Survey," Springer-Verlag, Berlin, Germany, 1981.
4. Yoon, K. Plau, Hwang, Ching-Lai, Multiple Attribute Decision Making: An Introduction, Series/ Number 07-104, 1995, Sage Publications Inc., Thousand Oaks, California, 1995.
5. Cassone, Deandra T., Tillman, Frank A., *A Professional's Guide to Problem Solving with Decision Science: An Integrated Approach for Assessing Issues, Finding Solutions, and Reaching Corporate Objectives*, Second Edition Overland Park, KS, Pioneering Partnerships, LLC, January 2018.
6. Tillman, Frank A., Cassone, Deandra T., *A Professional's Guide to Decision Science and Problem Solving: An Integrated Approach for Assessing Issues, Finding Solutions, and Reaching Corporate Objectives*, Upper Saddle River, NJ, Financial Times Press/Pearson Education, Web: December 1, 2012, Hardcover: February 2012.

Selecting the Best Method Supported by the Data

Abstract

Three fundamentals associated with robust problem solving and quantitative decision-making are data analysis, quantitative problem modeling and the inclusion of decision-maker inputs into the overall decision process. A problem solution approach that integrates these three elements captures key considerations that you will need to develop robust, quantitative decision-making processes and solutions. Several models that were developed using this approach are discussed in this section. These examples include studies that were conducted while working as a consultant and patents that were developed while working at a Fortune 100 company. Additional study references are also provided. These examples provide a general description of the problem, the data used in the model, the quantitative methods employed in the process and the decision-making techniques and are intended to show how this process works together to support quantitative decision making.

Introduction

It is fundamental that the decision maker understands how a quantitative decision-making process works to feel comfortable with decisions and results generated by the approach. The future of the company may depend upon these decisions. Thus, it is necessary to keep the decision-making processes understandable, and explainable. The simpler the method selected the better. For example, use a simple ranking rather than a complex model such as linear programming because the mathematics behind the ranking model are simple and easily understood.

D. Cassone, *Operational Leadership Using Quantitative Decision Making*,
Synthesis Lectures on Engineering, Science, and Technology,
https://doi.org/10.1007/978-3-031-67565-2_7

The model should be dynamic and easily adapted to fast-changing business conditions. The methodology selected should utilize readily available data and information. Some models are developed that require a significant effort to collect the data and maintain the model. If this occurs, more time is spent updating the data and the model rather than using the model to evaluate decisions. If it is too difficult to maintain a model, the model will quickly be forgotten.

The complexity of the business environment, the objectives, the availability of data, and the time frame for the decisions dictate what methods are best used to solve problems. Some models may require multiple criteria methodologies for their solution. Others may require incorporating several methods such as multiple objective decision making, statistical analysis, simulation, and expert systems. The goals and objectives of the decision makers can eventually point to the techniques to be used and the structure of the modeling process [1, 2].

The Three Key Components Associated with Problem Solution Development

The three fundamentals associated with robust problem solving and quantitative decision-making are data analysis, quantitative problem modeling and the integration of decision-makers into the overall decision process. Each of these three areas were discussed in the preceding three chapters. How then, do you bring together all of this information into a problem solution? To do so, it requires knowledge in the domain area of the issue, an understanding of the data available to solve the problem, knowledge and selection of applicable quantitative methods, and a knowledge of how to integrate decision-makers' goals into the problem solution. A solution approach that integrates these three elements captures key considerations that you will need to develop robust, quantitative decision-making processes.

Overall, the process described ensures that decisions that are made at the decision levels support the overall goals and objectives. Each of the steps in this process provide definition, structure and analysis that is beneficial in developing a model to represent the decisions being made as well as providing those doing the analysis with insight into the overall company objectives and company operations. Much information can be gleaned from this process.

Integrating all the data and processing computations together provides the user with a decision model to support the decision process. The models developed and used should be automated to the highest degree that is feasible and beneficial in its overall use. For example, a small model that is used to make decisions on a local level should not be automated in the same way as a decision application that is accessible to an entire company.

A key factor in automating a model to a certain degree is the ability to perform sensitivity analysis with the model. Generating the first answer from a model is only the start of the decision-making process. Various parameters and scenarios should be tested in the decision process to determine the best current and ongoing solution. Updating the data when required will keep the model current and representative of the current operating environment. As the environment changes, the model should be dynamic to adapt to the changing environment.

Modeling Considerations

To be most effective with quantitative decision-making, the team should develop a user-friendly system reflective of the technology that allows decision makers interaction with the decision process from problem definition to solution generation. Most decisions faced by executives are based on a number of conditions, goals, and objectives which must be satisfied, and are usually conflicting. With this in mind, the team should approach decision modeling by incorporating experience and expert judgement along with statistical data to solve multiple criteria problems and perform complex trade-off analysis to facilitate individual and group decisions. It sounds like a tall order of business, but examples will be provided later in this chapter to demonstrate these concepts.

To form a firm foundation, those developing and applying the decision model should be versed theoretical methods and application of methods in model development and have the ability to evaluate theoretical approaches and select and apply only those that are best suited to solve practical problems. The "best approach" is the one that will be used and the "optimal solution" simply interprets to "are you doing it any better than you were." The models must be dynamic and must be easily changed to meet the needs of a constantly changing environment and that experience and expert judgement must be incorporated into the models along with "hard" data [1, 2]. Key considerations in this process include:

- The decision maker's interaction with all phases of the decision process.
- Integration of input from several decision makers into a consensus decision.
- The time frame available for the construction of the model.
- The complexity of the environment to be modeled.
- The need for the model to be dynamic.

There is no silver bullet in developing quantitative decision-making models. The concepts presented in this book provide an investigative framework that can assist you in assessing your own specific environment and developing approaches to problems that meet your unique environment and needs. Below are some real-world problems that have been solved with this approach and insights into how each of the elements described in the previous chapter have been used and applied to solve these problems.

Model Development Examples

This section discusses studies and patents that used the concepts in quantitative decision-making to solve problems for the military and industry. These studies and patents include understanding the environment, gathering decision-maker's objectives, analyzing the available data and determining additional data needed to solve the problem, employing quantitative methods and applying decision methodologies were all used to develop solutions. These concepts were integrated together to form robust solutions to crucial problems.

You will see an overlap in quantitative methods that are used to develop the resulting models and the incorporation of the decision maker in the modeling process. As discussed in previous chapters, there are quantitative decision-making methods that can be built into the decision-making process. These methods serve as both a quantitative method and an insurance that the decision maker is incorporated into the decision process. The overlap of methods and decision-making methods is helpful to ensure that decision makers are well integrated into the decision process.

Investment Strategy for Military Command

Problem Definition

An investment strategy and supporting process was developed for a military command that does research and development for the military aerospace industry. This is an interesting study in that there was a diverse set of projects presented to the executive team for funding and required the development of a set of goals, objectives and metrics that could equally represent the projects and compare them. The end product of the study was a list of projects that would be funded based on various budget constraints.

This approach established goals and objectives for the military command's research projects that could be used to analytically prioritize programs and allocate resources for the military projects. This approach required that senior executives critically examine their goals and objectives and their relative importance in the final decision. The decision criteria and measures were then developed to provide a means to evaluate projects. Once evaluated, the projects were then ranked using these decision criteria and measures and their relative importance to allocate resources based on the goals and objectives. The challenge to ranking the projects lied in the wide range of activities covered and to determine a set of that is applicable across the wide range of activities.

Group decision making techniques were used to provide a means for developing the goals, objectives and decision criteria used in the decision process and weighting the importance of the objectives and criteria. All the decision makers had input in the final decision process insuring a greater level of objectivity in the project evaluation process. The various viewpoints of the members of the executive directors and others were captured, and a group consensus/tolerance achieved in a set of decision criteria. The investment strategy developed in this process provided the foundation for establishing an objective, automated structured project prioritization and resource allocation process that provided the command with a traceable and defensible basis for investment decisions. The steps and reports were specified so that an automated system could be developed to efficiently repeat the process in the future [3].

Data Mining and Statistical Analysis

Various techniques such as surveys, group discussions, working group meetings, individual review and feedback consolidation were used to capture and assess relevant data for the decision-making process and arrive at the resulting list of the military's goals, decision criteria, measures and resource allocation constraints. This information formed the basis for the project evaluations that were to be evaluated by the decision-making model. Each research project was then evaluated using objective or subjective data to provide input into the decision-making model.

Quantitative Methods

This project utilized data analytics, multiple attribute decision making, group decision making and resource allocation methods to understand key pieces of information required for the analysis, capture the appropriate data to rank and to allocate resources based on the application of multiple attribute decision making techniques.

Applying Decision Methods

This study relied heavily on group decision making and multiple attribute decision making techniques. From a group decision making perspective, the team used brainstorming techniques, surveys, a version of the Nominal Group Technique (NGT) and feedback mechanisms to gain a group consensus of the goals, objectives, decision criteria and metrics that would be used in an evaluation process. Significant work and consensus building was required to develop the finalized set of objectives and decision criteria. Additionally, all of the executive committee members wanted a direct say in the importance of the objectives and criteria and the individual evaluation of the projects. Group consensus techniques were used to develop the model criteria weights and to aggregate the input values of the executive committee members for each project evaluation into one group consensus evaluation score so that it could be run through the multiple attribute decision making process. Tables 1, 2 and 3 show a summary of the resulting goals, decision criteria, descriptions and measures for each of the goals.

The multiple attribute decision making model used the Simple Additive Weights method to prioritize the projects. All scoring was done on a 1–10 scale so that normalization was not required from the input data. The model was structured hierarchically where there were overarching objectives that were weighted and then individual criteria

Table 1 Goals, decision criteria descriptions and measures for Goal 1

Goals/decision criteria	Description	Measure
Goal 1. Satisfy customer needs (Major Commands, Operating Bases, T&E, Intelligence, Aerospace Industry)		1.a–1. d
1.a Documentable customer need	A user need which is documentable and identified.	Range of 0–10 where 0 = neither are specified 10 = both are clearly specified
1.b Broad applicability	Broad refers to either broad range of customers, broad range of specific applications or broad range of systems	Range of 0–10 where 0 = no direct application 10 = pervasive
1.c Importance of need to the military	Importance refers to the priority of the need to the military	Range of 0–10 where 0 = not demonstrated 5 = enhances a system 10 = critical or enabling for system *Enhancement* is an improvement to performance. *Critical* is a technology needed to meet system requirements. *Enabling* is a technology needed to do a new mission.
1.d Breakthrough	Breakthrough technology that could provide a profound impact on a mission capability	Range of 0–10 where 0 = evolutionary or enhancement 5 = leaps to state-of-the-art 10 = provides a unique capability

within the objectives that were also weighted to arrive at a resulting weight for each of the unique decision criteria.

Resource allocation methods were also used to allocate the budgets across the projects. There were multiple sources of money and multiple conditions associated with the resource allocation process, such as "each functional area will have at least one project funded" and they were accounted for in the resource allocation process. This was done manually, due to the limited number of projects that were prioritized. This additional information was used along with the project prioritization so that they met their overall strategic mix of projects. A multiple objective decision-making model could have also been used to allocate resources.

Ultimately, this study used a number of decision-making tools and methods to ensure that the perspectives of the leaders were integrated into the decision process. NGT, surveys

Table 2 Goals, decision criteria descriptions and measures for Goal 2

Goals/decision criteria	Description	Measure
Goal 2. Make a difference for the military in this technical area		2.a–2.e
2.a Leadership	Can we provide the vision or set future direction? Do we have a well-defined niche and the personnel to lead the effort?	Range 0–10 where 0 = no unique role for us 10 = only we can meet this need
2.b Adequate funding	Adequacy of the proposed funding.	Range of 0–10 where 0 = unrealistic to do the task 5 = adequate to meet minimum requirements 10 = perfect fit of funding to task
2.c Adequate work years	Adequacy of the proposed work years and whether the right people are proposed for performing or managing the work	Range of 0–10 where 0 = unrealistic to do the task 5 = adequate to meet minimum requirements 10 = perfect fit of work years/people to task
2.d Payoff	The potential technological payoff of the development effort.	Range of 0–10 where 0 = low 5 = moderate 10 = high
2.e Soundness of technical approach	This should include whether or not the approach is scientifically valid, technically supported and logical	Range of 0–10 where 0 = unreasonable and unsupported 10 = sound and well thought out

and meetings were used to gather input and information from the key executive decision makers. This input was then used to develop a multiple attribute decision making model that could be used to prioritize the projects across a diverse group. Integrated together, the group then generated the data to evaluate the projects across the directorate and develop a group consensus evaluation for each of the projects. Then, these ranked projects were evaluated against the available dollars and the budgeting nuances and constraints to determine which projects would be funded and pursued. The Investment Strategy prioritization process provided the foundation for establishing an objective, structured project prioritization and resource allocation process which provided a traceable and defensible basis for investment decisions.

Table 3 Goals, decision criteria descriptions and measures for Goal 3.

Goals/decision criteria	Description	Measure
Goal 3. Have a transition strategy/technology plan		3.a–3.c
3.a Customer flow	All customers excluded previously are included. Logical flow of customer to customer planned. Partnerships or commitments established. Specific end use requirements identified	Range 0–10 where 0 = plan is unworkable 10 = good progression of technology
3.b Risk of transition strategy/technology plan	Components of risk include •All key technical issues are addressed • Schedule meets time frame of customer needs •Schedule realistically allows enough time for technology development •Major milestones for each stage of development are defined and measurable •Supporting technology demonstrations are in place in time	Range 0–10 where 0 = risk is unacceptable 10 = risk is acceptable considering the potential payoff
3.c Leverage of other resources	Resources external to normal resources are being used to accomplish development and transition/transfer of this technology. Customer or other government organization are willing to provide some transition funding	Range of 0–10 where 0 = no attempt to leverage resources 10 = outstanding use of other resources

Logistics Pipeline Analyzer

Problem Definition

Performance tracking and analysis provided users at all levels of the organization with the ability to better manage the logistics pipeline. Significant resources are expended in materiel management for the Department of Defense (DoD). The identification and improvement of any and all of the pipelines and pipeline segments provided significant opportunities in cost reductions and improved customer service. The Logistics Pipeline Analyzer (LPA) was designed to provide users at all levels with the ability to analyze and assess logistics performance at each of the stages of logistics management.

By clearly understanding current performance, historical performance and performance trends, management could focus efforts on areas where performance can be improved.

One of the key characteristics of the LPA system was its flexibility. The LPA system was designed to accommodate management and management analysis at all levels. Performance summary information could be rolled up, for example, for a subordinate command, or could be analyzed on an individual National Stock Number (NSN) by NSN basis.

Another important characteristic of the LPA system was its ability to support the detail that was used to generate the various system performance calculations. Certain performance statistics could be generated for a selected group of NSNs. Results may have appeared reasonable or appear to be out of the ordinary. The LPA system provided the back-up detail that was used to support the results generated by the system. Many times, a "bottom line" performance statistic may not tell the entire story. In some cases where lead times may appear to be high, the backup detail showed that extenuating circumstances surrounded a specific NSN. Other times, there may be no extenuating circumstances, and the conclusion were that this is an area that needs additional attention.

Providing the ability for all levels to view and analyze performance aided in identifying performance improvement opportunities. Business process enhancements could be identified and suggested so to reduce the resource investment in the pipeline by reducing the time and size of the pipeline functions. The LPA system provided the analysis capabilities to better manage cost and monitor performance leading to increased responsiveness to customer requirements and improvements in the bottom-line cost of logistics operations. Figures 1 and 2 below show examples of the Logistics Pipeline Analyzer system [4].

Data Mining and Statistical Analysis

This study relied heavily on data and statistical analysis. The premise of the system was a data driven representation of the DoD's logistics pipeline and tracked and analyzed data that supported the major weapon systems to an individual part or NSN level. The definition of requirements and time that it took to develop this system was significant. In general, the following steps were used to develop the metrics-based system.

Steps to Lay Out the Metrics-based Performance Measurement System

- Identify major logistics performance concerns for each level of management.
- Identify products/results associated with each major logistics performance concerns.
- State specific goals/objectives in each process/results/product area.
- Identify appropriate measures and data sources for objectives.
- Evaluate measures in terms of norms, baselines, trends.

This process involved starting with some reasonable set of metrics, assessing the usefulness of the metrics in measuring performance and creating, over time, an optimally useful set of metrics.

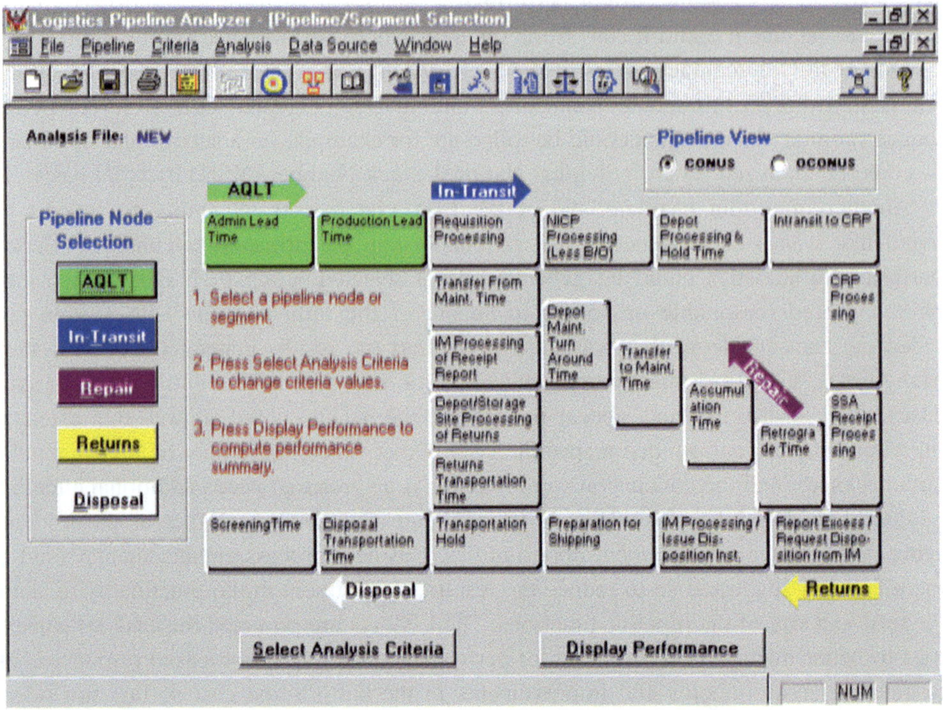

Fig. 1 Logistics pipeline analyzer

Quantitative Methods

The nature of the LPA system was such that it was a quantitative decision-making system. The logistics performance management system provided the ability to perform performance tracking using two primary representations of the product/part data. The first performance statistics tracks the lead time associated with a specific product/part. The second performance statistics tracks the dollar investment value of a product/part as associated with the lead times and the item costs. Any of the defined pipelines or pipeline segments could be selected for the analysis. This information was grouped using any variety of criteria including cost centers, departments, priorities, part types and sources of supply. This information was maintained on a time series basis such that a specific period of time can be viewed for the selected grouping of product/parts.

Examples of some of basic performance analysis capabilities for performance tracking for individual or a group of product/parts are listed below:

- On a daily, weekly, monthly, quarterly or yearly basis
- By cost centers, departments, or supply source

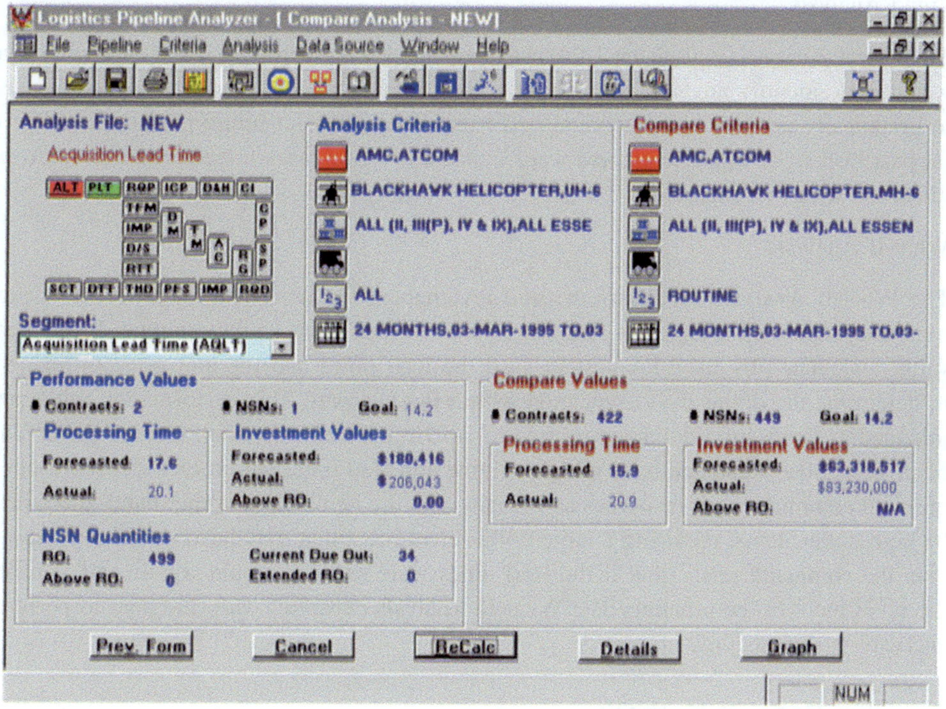

Fig. 2 Logistics pipeline analyzer analytical capabilities

- Based on specific characteristics associated with a product/part such as type of part, priority, product material, or shelf life.
- Performance compared to Corporate Goals

These types of capabilities provided the user with the ability to analyze and manage areas that they could directly influence. The specific analyses performed enabled the user to identify and highlight opportunity areas for improvement. The analysis capabilities discussed below provided additional techniques available to further analyze the product/part data tracked with the system.

Compare Analysis

A compare analysis provided management at the various levels of the organization with the ability to compare the performance of one grouping of products/parts against another grouping of or individual products/part(s). The analytical capabilities were the same as those provided with the basic performance analysis, however, this capability can be used to perform "side by side" comparisons of the various cases.

Trend Analysis

The Trend Analysis within the LPA provided another powerful analysis capability that was used to identify areas of potential operating improvements. Trend Analysis was used to graphically analyze past historical performance and project future performance based on past history. Areas where there were increases or decreases in performance could be identified with this capability.

What-If Analysis

The What-If Analysis capability of the LPA enabled the user with the ability to determine the impact of changes in lead times on the investment in the pipeline. Investment values calculated by the LPA system used the lead times and the unit price of an item to determine the dollar-days associated with a Product/part. This performance value was intended to represent a dollar figure tied to DoD resource investment. Certain items may be of low dollar value and have a high volume and lead time, but, in essence, represent a minimal commitment of resources associated with the item. On the other hand, there may be high dollar items, requiring a large dollar investment that may have lead times shorter than the command goals, that if the lead times were reduced, could significantly impact the investment in the pipeline. The What-If Analysis capability was designed to perform an analysis such as this.

Applying Decision Methods

This system was developed as a metrics-based system. Decision makers at all levels of the organization were included in the development of the metrics based on the performance. This information was integrated into the overall system implementation. Decision makers were surveyed to identify major logistics performance concerns for each level of management, identify products/results associated with each major logistics performance concerns, stating specific goals/objectives in each process/results/product area, and identifying the appropriate measures and data sources for the objectives. Group decision making methods were used in the development of the ultimate set of weapon systems and metrics that were integrated into the automated system.

The LPA results presentation was formulated to facilitate the decision-making process. Reports and graphs were generated by the LPA for any of the above analysis. This provided the user with the ability to present findings to those at higher and lower levels of the organization to highlight opportunity areas and present justifications for the finding of the analysis. At a lower level of the organization, the user provided the results of an analysis and ask subordinate managers for suggestions for opportunities for improvements or rationale for current performance. At higher levels of the organization, the results from the LPA were used to suggest and justify areas that by making policy or operating changes can improve the bottom-line performance of the pipeline or pipeline segment. The presentation of the data and findings from analyses generated by the LPA were used to increase understanding of the pipeline operations as a whole and initiate changes that optimized the time, resources, and dollar investment associated with logistics.

Organizational Capacity Model

Problem Definition

A manpower planning decision model was developed for a telecommunications company to determine which projects were of greatest importance to accomplishing the business goals of the organization. This model helped determine the appropriate skill mix of individuals that was required to complete these projects. Primary steps in the development process were developing criteria to evaluate the projects being worked on by the organization, defining the skill categories and skill levels needed for each project, defining the work units within the system, and formulating the decision model into an easy-to-use system that could adapt to their quickly changing environment [4]. Figure 3 shows a flow diagram of the process.

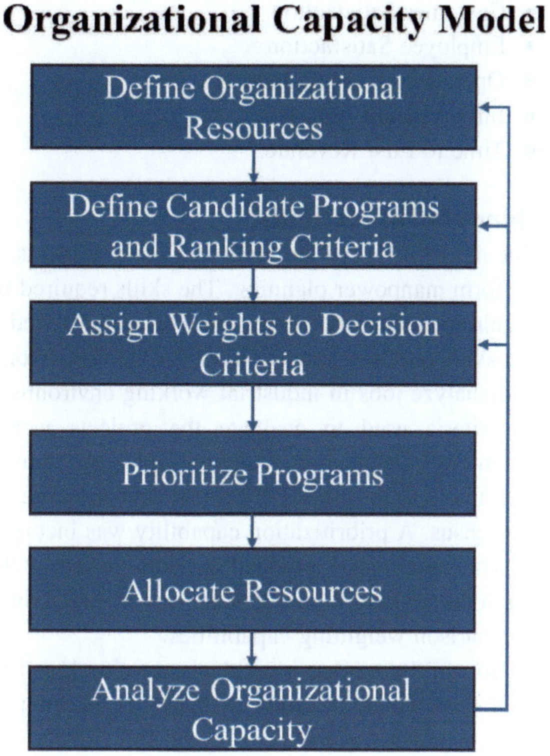

Fig. 3 Flow diagram of the organizational capacity model

Data Mining and Statistical Analysis

Within the organization, fifty-six skills were identified that were required for the work and five levels of skill could be utilized. With the system these skills were reduced to several broad categories such as engineer, management, programmer, or specialist which simplified the modeling and data requirements. The work units of staff months were defined in terms of resources and skills required for completing each of the projects. A skill level assignment was also made for the special skills required on the project. Resources and timing were also a part of the requirements to be satisfied before a project was undertaken.

Based on discussions with executives and a group consensus building activity, each project was evaluated against an agreed upon set of decision criteria and associated metrics and was then prioritized to determine the importance of the project in achieving the overall objectives of the organization. The criteria used for the prioritization included:

- Return on Equity Impact
- Profit Growth Contribution
- Market Share Impact
- Customer Satisfaction
- Employee Satisfaction
- Operations Cost Reduction
- Infrastructure Impact
- Time to First Revenue.

Quantitative Methods

The model included the ability to assess projects, assign personnel to projects as well and perform manpower planning. The skills required for a project were matched to the project requirements and resources were allocated based on timing, availability of resources and the skills required and available for the task. Work design concepts were used to evaluate and analyze jobs in industrial working environments.

Criteria used to evaluate the projects were defined, the metrics and evaluation parameters (subjective scales or objective numbers) decided upon, and the data input structure was developed. Pertinent decision criteria were developed based on management consensus. A prioritization capability was incorporated into the model.

The system also provided decision makers with the ability to weight the decision criteria in the ranking. Criteria weighting included direct (normalized) weighting and pairwise comparison weighting capabilities.

To run the model, individuals were assigned to projects based on a general technical skill type which included manager, engineer, programmer, or specialist. Each of these types also had several skill levels. Individuals were then assigned to the project by skill type and skill level. The number of staff months that the individual had available was matched with what was required to perform the project.

A staff month budget allocation was made for each project which included the minimum level of staff hours required for the project and the total hours required for the project. The project could not be performed if the minimum available hours were not enough to do the project.

Projects were then evaluated against the criteria and a ranking was performed. A resource allocation was performed in terms of staff hours including value-based budgeting. Decisions were made as to what projects would be completed or not completed by the company. This then drove the skill type and number of personnel that would be required for the completion of this work within a time limit.

Applying Decision Methods

This model incorporated two facets of corporate decision-making. The first being the types of projects that the organization undertook to support the long-term objectives of the organization. This was handled by incorporating a multiple attribute decision making model into the system. The second was types of individuals that were required to develop the projects with regards to manpower planning. The corporation could then look at the types of individuals that were available by skill type and skill level to decide whether additional or fewer individuals would be required to complete the workload. This was handled by incorporating a resource allocation capability into the system. This process focused on the types of individuals required to meet their project base workload while accomplishing the goals of the organization.

Throughout this process, group decision making techniques were used to establish the decision criteria used in the prioritization process. Group decision making techniques were also used to identify the fifty-six job classifications and then reduce those skills to a manageable set of skills for the decision-making models.

Additional Studies for Review

A number of additional examples employing the basic methodology discussed in this book are presented in additional studies and books by the author and Dr. Tillman. This includes studies such as those for major corporations for research and development prioritization activities [5], the development of a manpower planning model [6] and the development of a Warehouse and Inventory Level Optimization model for a brewing company [7]. Each of the steps in the process flow described in this book are not explicitly called out, however, the methodology, in general, described in this book was used to perform these studies.

Quantitative Decision Making in Patent Development

While working at a telecommunications firm, the author developed and was ultimately awarded a number of business process patents. This section contains examples of those patents. The fundamentals of the approach described in this book drove the development and award of these patents. The author believes that applying the thought process that is presented in this book and was used in consulting projects led to the development of unique processes that could then be submitted as patents. The knowledge of methods and tools was also very beneficial. Methods were used in traditional and used in non-traditional ways led to innovation.

Project Predictor and Enhanced Project Predictor

The project predictor patent involved developing a quantitative decision-making model that could be used in the Project Management Office to predict the duration of a project. The development of this system involved capturing historical data about the operating environment, attending core team and planning meetings and learning about key factors that impacted the project cycle time. Fuzzy Logic was used to find the most robust model for the problem solution. A second patent was filed and included enhanced variables in the model [8, 9].

Some fairly traditional historical data was used to formulate the model including various in-house and vendor cost estimates, the type of release that the project would be launched, the number of applications that the project would impact and other factors. The more unique variables that were included in the model were derived from the core team meetings that were attended. Individuals said consistently that if network and/or IT development was involved in the project, the project took longer to complete. They also said that the larger the core team, the longer the project took to complete. Because of these real-world operating characteristics, Boolean variables were used to identify the organizations part of the core team and the total number of core team members were summed. These variables were included in the stepwise regression analysis that was used to generate the prediction. took longer time.

Also, when the historical data of time vs project cost was plotted, it showed a definite difference in the data at $1.5 million dollar project costs. When projects cost below $1.5 million dollars, a line could be well fit to the data. When the project costs were above $1.5 million dollars a separate line could also be well fitted to the data. This resulted in developing two regression models, based on the total estimated project cost. Figure 4 shows an example plot of the data.

Of interest also, is that the key decision makers in the Project Management Office were involved in the development, oversight, and acceptance of the model. These decision makers could verify from real-world experience that the characteristics of the environment were represented in the model and that the model produced realistic results.

The fuzzy logic component of the model was used to assess the model's performance. Membership functions were developed and aggregated to assess which model was the best performing model.

Overall, the patents included an understanding of the environment, engagement of the decision makers, data analysis, quantitative methods, model automation and verification. Also, the model was automated with a database and user front-end so that individuals and

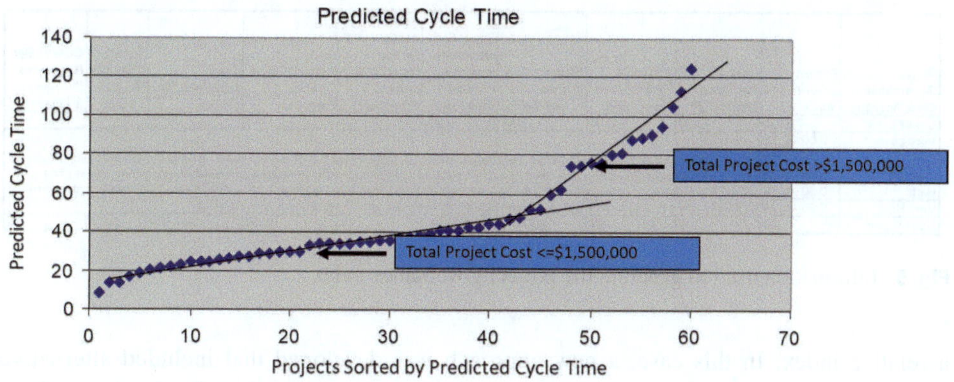

Fig. 4 Plot of the cycle time data

decision makers could easily interact with the model. The model was run at the beginning of each project to then provide an estimate of the overall cycle time or duration of the project.

Flexible Project Governance

This patent provided a flexible governance model that is used to monitor project progress based on project difficulty and project performance criteria [10]. The Project Management Office at the time employed a stage gate process for assessing project performance. Project managers would present their project status at the completion of a project phase to a committee of approximately fifteen cross-functional corporate executives. Project managers had templates that they would complete that addressed key aspects of the project such as the planned to actual time and costs, status on the project scope and any other issues associated with the project. Since this was a high-visibility presentation, the project managers would spend a significant amount of time developing the presentations for the review. Some individuals within the company viewed this as unnecessary if the project was performing well, while other times, projects that were performing poorly may need to have additional reviews to address any issues and keep the project on track. To address this, a model was developed to predict project complexity of the project and to predict the project governance needs based on the project complexity and project performance.

Specific operational performance metrics were developed to represent project complexity attributes and compute a Project Complexity Index (PCI). A Project Governance Index (PGI) was also developed using key project performance characteristics to determine the initial and then ongoing project governance needs based on performance.

The Project Complexity Index (PCI) was generated to evaluate project complexity factors and assign an overall project complexity level. The Multiple Attribute Decision Making method called TOPSIS (Technique for Order Preference by Similarity to Ideal Solution) was used to generate the index. The traditional TOPSIS methodology generates

Project Governance Classifications	Project Governance Index	Cost Performance Index - CPI	Schedule Performance Index - SPI	Elapsed time in phase (% of total time)	Expended resources in phase (% total resources)	Scope assessment (scope reductions 1 - none to 5 significant)	Non-Administrative Change Requests	Team Turnover (% churn)	Sponsor Change (0 - no to 1 - yes)	Constrained Resources (1 - none to 5 - many)
Lower Limit	0.00	0.00	0.00	0.0%	0.0%	1	0	0.0%	0	1
Very Low	0.00 - 0.19									
Low	0.20 - 0.39	This complexity index is used to generate one score to represent project governance requirements. A project is scored using these								
Medium	0.40 - 0.59	criteria. A Multiple Attribute Decision Making method such as TOPSIS is used to generate a complexity index for the project. The								
High	0.60 - 0.79	project governance index falls between the upper and lower limits and is in the range of 0.00 to 1.00. The resulting index can be kept as								
Very High	0.80 - 0.99	an index score or put into governance categories such as Very Low, Low, Medium, High and Very High.								
Upper Limit	1.00	2	20	200.0%	500.0%	5	200	200.0%	1	5

Fig. 5 Information used to generate the project governance index

a relative index. In this case, a new approach was developed that included alternatives with the upper and lower bound of the possible outcomes "force" an absolute, not relative index. This is used with the PCI as well as the PGI and also used in the patent Systems and Methods for Identifying High Complexity Projects [11] to identify high complexity projects so that they can be broken down into smaller project with a greater chance for successful completion. Attributes could also be weighted with the absolute TOPSIS method, or all have equal weighting. Project attributes included the PCI, vendor data, alpha and beta testing needs, network, IT and device impacts, project dependencies, new technologies, project cost and other criteria. The PCI was used to determine the initial governance structure for the project. Figure 5 shows an example of the information used to generate the PGI.

The Project Governance Index (PGI) used project health characteristics to generate an index that represents whether or not the project was performing as it should. This index also used TOPSIS with an absolute index so that a consistent measure could be generated in absolute terms instead of relative project terms. Data included in the computation of the PGI included cost, schedule and scope assessments, elapsed time and resources expended, change requests, team turnover, sponsor changes and resource constraints. The PGI mapped various governance actions on a periodic basis (monthly) and more or less project reviews were deemed necessary based on the computation of the index. Figure 6 shows the real time project intervention used to manage the projects.

Historical and project assessment data were used in the model. Key decision makers agreed upon the data that would be used to evaluate the projects and the framework for more or less reviews of the project by the executive team. The development of this patent involved understanding the environment, engaging the decision makers, assessing and developing data inputs, engaging decision makers in the project review changes and developing a spreadsheet application to use the approach on an ongoing basis.

Additional Patents

Additional patents that have been awarded based on the approach provided in this book are briefly mentioned below. These patents span functions such as project management,

Fig. 6 Real time project intervention

Real Time Project Intervention

PGI	Governance	Stoplight
0.6+ NA		Green
0.5-0.6	Alert Only	Yellow
0.4-0.5	Mgr Review	Yellow
0.3-0.4	Director Review	Yellow
0.2-0.3	VP Review	Red
	Interim Gate Review	
0.0-0.2	Gate Review	Red

order management, marketing, customer care, inventory management, order fulfillment, and shipping.

- Risk Management in Project Management—This patent involves creating a risk management workbook that can be used along with project management techniques to mitigate risk [4, 12].
- Problem Ticket Cause Allocation—This is a control system that uses the concept of root cause clusters to identify and alert individuals of potential order fulfillment, ordering system and operational issues [5, 13].
- Habituation Oriented Product Selection System—This patent uses the concept of habituation to select new telecommunication devices to launch [6, 14].
- System and Method for Care Complexity Model—This patent is a model that is used to measure and monitor the volume and complexity of information provided to call center representatives [7, 15].
- Inventory Variance Detection Modeling—This patent involves analyzing non-traditional systems and processes to identify upstream and physical causes associated with inventory variances in a warehouse [8, 16].
- Inventory Loss Prediction—This patent uses statistical modeling and analytical techniques to predict telecommunication device losses identified by internal and external customers [9, 17].
- Order Fulfillment Logistics Workflow—This patent applies the concepts of flexible manufacturing to warehousing to consolidate warehouses and adapt the warehouse to meet changing customer demand [10, 18].
- Shipping Aggregation in Mobile Phone Order Fulfillment—This patent is a repeatable process that is used to identify opportunities for order aggregation in the warehouse to reduce shipping costs in the telecommunication industry [11, 19].

References

1. Cassone, Deandra T., Tillman, Frank A., *A Professional's Guide to Problem Solving with Decision Science: An Integrated Approach for Assessing Issues, Finding Solutions, and Reaching Corporate Objectives*, Second Edition Overland Park, KS, Pioneering Partnerships, LLC, January 2018.
2. Tillman, Frank A., Cassone, Deandra T., *A Professional's Guide to Decision Science and Problem Solving: An Integrated Approach for Assessing Issues, Finding Solutions, and Reaching Corporate Objectives*, Upper Saddle River, NJ, Financial Times Press/Pearson Education, Web: December 1, 2012, Hardcover: February 2012.
3. Tillman, Frank A., Cassone, Deandra T., Investment Strategy for Product Development in the Aerospace Industry, Financial Times Press/Pearson Education, Upper Saddle River, NJ, Web. May 31, 2012.
4. Cassone, Deandra T., Tillman, Frank A., The Science of Common Sense: Best Practical Decision Science Methods, HTX, Inc., Osage Beach, MO, October 13, 2015.
5. Tillman, Frank A., Cassone, Deandra T., Strategic Planning and New Product Development, Financial Times Press/Pearson Education, Upper Saddle River, NJ, Web. May 31, 2012.
6. Tillman, Frank A., Cassone, Deandra T., Manpower Requirements for Management and Professional Personnel, Financial Times Press/Pearson Education, Upper Saddle River, NJ, Web. May 18, 2012.
7. Tillman, Frank A., Cassone, Deandra T., Developing a Warehouse and Inventory Level Optimization System, Financial Times Press/Pearson Education, Upper Saddle River, NJ, Web. May 31, 2012.
8. Cassone, Deandra T., et.al, Project Predictor, U.S. Patent No. 7,676,490 (issued March 9, 2010).
9. Cassone, Deandra T., et.al, Enhanced Project Predictor, U.S. Patent No. 7,949,663 (issued May 24, 2011).
10. Cassone, Deandra T., et.al, Flexible Project Governance based on Predictive Analysis, U.S. Patent No. 8,108,238 (issued January 31, 2012).
11. Cassone, Deandra T., et.al., Systems and Methods for Identifying High Complexity Projects, U.S. Patent No. 7,983,946 (issued July 19, 2011).
12. Cassone, Deandra T., et.al., Risk Management in Project Management, U.S. Patent No. 8,005,706 (issued August 23, 2011).
13. Cassone, Deandra T., et.al., Problem Ticket Cause Allocation, U.S. Patent No. 8,028,197 (issued September 27, 2011).
14. Cassone, Deandra T., et.al., Habituation Oriented Product Selection System, U.S. Patent No. 8,055,546 (issued November 8, 2011).
15. Cassone, Deandra T., et.al., System and Method for Care Complexity Model, U.S. Patent No 8,155,996 (issued April 10, 2012).
16. Cassone, Deandra T., et.al., Inventory Variance Detection Modeling, U.S. Patent No 8,249956 (issued Aug, 21 2012).
17. Cassone, Deandra T., et.al., Inventory Loss Prediction, U.S. Patent No. 8,332285 (issued December 12, 2012.

18. Cassone, Deandra T., et.al., Order Fulfillment Logistics Workflow, U.S. Patent No. 8,644,983 (issued Feb 4, 2014).
19. Cassone, Deandra T., et.al., Shipping Aggregation in Mobile Phone Order Fulfillment, U.S. Patent No. 8,756,119 B1 (issued Jun 17, 2014.

Implementation

Part III addresses modeling the developed solution and decision-making process as well as monitoring the success of accomplishing the objectives. This part provides a high-level discussion of putting the data, methods and decision-making models into practice. Automating the processes leads to success and ultimately, you will want to monitor and measure the accomplishment of the objectives. (See Fig. 1)

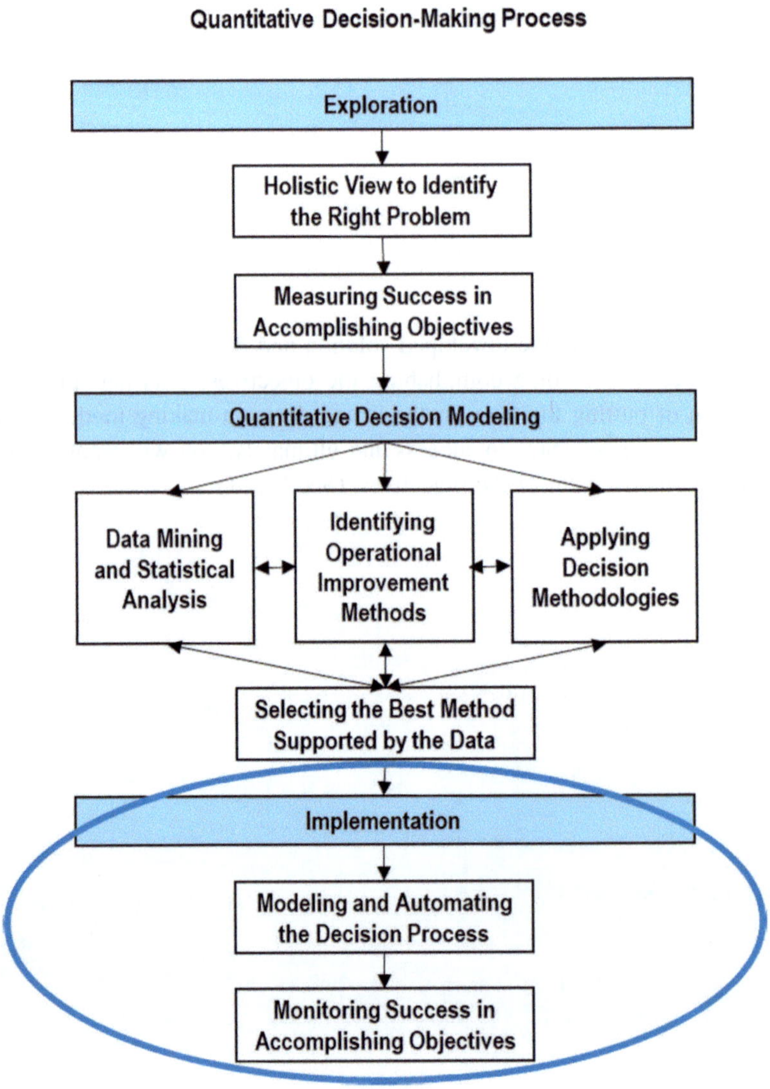

Fig. 1 Implementation

Modeling and Automating the Decision Process

Abstract

This chapter discusses different ways to implement your quantitative decision-making projects and solutions. Formalized projects, identifying "Quick Hits," or large scaled efforts may be required to move your organization towards quantitative decision-making processes. Project scoping and definition is discussed along with engaging appropriate stakeholders and developing project proposals. Different automation options are also provided so that you can scope your operational model to the nature, frequency, and use of the model in quantitative decision making. Business case development may also be required to provide financial justification to the organization for implementing projects. Project and life cycle costing considerations are also addressed.

Identifying Projects

As you use this process, you may identify a formalized project that is required to support your quantitative decision-making process. These projects can take many forms such as those that can be easily implemented on the "factory floor" or projects that will require a level of system scoping and development to support the process. The author has very broadly divided these projects into two categories. Quick Hits are those projects that can be easily implemented. Larger scale projects typically employ organizational resources such as time and money and require a more formalized definition of the work to be accomplished, an approval to proceed and a use of greater organizational resources to accomplish the project. For larger projects, project management processes are typically used to implement these projects along with a justifiable business case to determine the return on the effort. Each of these types of projects are discussed below.

D. Cassone, *Operational Leadership Using Quantitative Decision Making*,
Synthesis Lectures on Engineering, Science, and Technology,
https://doi.org/10.1007/978-3-031-67565-2_8

As a note, there is overlap in the following section in "Selecting the Best Method to Support the Data" and this section, "Modeling and Automating the Decision Process." The previous chapter focused more on the method and models used in the studies. This chapter will focus more on the project development processes to implement the quantitative decision-making models.

Quick Hits

Quick Hits are those projects that are easily identified and relatively easily implemented. Typically, they may take the form of simple modifications to processes and the use of information that can be accomplished with a limited scope. The author was a lead in a large-scale end-to-end device quality process improvement effort at a national telecommunications firm. This project encompassed process mapping the entire device launch process, from device conceptualization to device delivery, to identify process improvements and improve device quality. In this effort, the author was asked to provide a list of Quick Hits to the senior executive supporting the project so that improvements could be implemented quickly. One of the Quick Hits identified as part of the study was to change the device quality inspection schedule to better align with the overall device launch schedule. This project was implemented quickly and simply was a scheduling change.

Another Quick Hit improvement that could be done with minimal resources and time involved the device prototyping process with device manufacturers. Typically, when the product and device development team met with the Original Equipment Manufacturers (OEMs) on a quarterly basis, they provided a bulleted list of five or six features that they would like to see on the new devices. Sometimes, the vendor came back a few months later with a device that met the product development team's requirements, other times the vendor was completely off in what the product development group wanted. When a fully designed device is sent to the manufacturer, typically a large number of design specifications are included in the design. The Quick Hit suggested for the concept development phase was to identify 20–25 key design specifications for the new device and provide that to the OEM to develop. This provided the OEM with much clearer direction with regards to the device design. The implementation process for this was working with the engineers to identify those key specifications and then get the approval from the senior executives that these were the key specifications to send to the OEM. This implementation took minimal time and resources but was impactful.

Larger Scaled Projects

Many times, projects that support the quantitative decision-making process require a greater use of resources and require formal project definition and approval processes. For example, if decision makers identify that there are significant issues in meeting customer demand and that a key element of the problem lies in the delivery scheduling of trucks to individual customers, the organization may want to review various route scheduling options that are available for their transportation fleet and implement a new transportation system. This type of project would require capturing the goals, objectives, decision criteria and metrics of the organization, including them in the decision-making process, evaluating a number of scheduling systems based on the organization's goals and then implementing the new scheduling system to better meet the needs of the organization. This type of effort would most likely require a formalized project with executive approval and project management support in the implementation.

Additionally, an organization may adopt a new way to evaluate their project and portfolio decisions by adopting an evaluation approach relying on quantitative decision making methods. Organizations may implement a multiple attribute and group decision making process that is used to evaluate product line-item deletion and retention decisions, organizational capacity and workforce decisions, research and development decisions, portfolio management decisions and others which required the development of an automated tool to support a robust decision-making process. The implementation of a quantitative decision-making system could also easily encompass the need for a more formalized project management approach. While Quick Hits can be implemented with minimal resources, larger scaled projects typically require formalized project management processes and are discussed in the following sections of this chapter.

Project Scoping and Definition

Project management, its key concepts and methods are an entire field of study and will not be discussed in detail in this book. Kerzner [1] book on project management provides a comprehensive resource for project management. Verzuh [2] book provides a practical and implementable approach to project management, and both are useful resources. Three key aspects in the acceptance of a formalized project include developing a project proposal as per the organizational requirements, clearly understanding and engaging the project decision makers and stakeholders and developing a robust business case to provide the financial and strategic support of the project. If the project is approved, a project plan will ultimately need to be developed. Each of these key aspects are discussed further in the following sections.

Project Proposal

A project proposal defines the future business value of the project. The proposal articulates these benefits in relationship to expenses for the project. Project proposals usually communicate to the evaluators and decision makers the desirability, viability, and feasibility of the project. A project proposal framework usually includes the following.

a. Project Team
b. Project Goal
c. Problem/Opportunity Definition
d. Proposed Solution
e. Project Selection & Ranking Criteria
f. Cost/Benefit Analysis
g. Business Requirements
h. Scope
i. Major Obstacles
j. Risks
k. Schedule Overview

Project proposals may be required by your organization in a specified format. If they are not required by your organization, the exercise of creating a project proposal is valuable to you as the project manager or project implementor. Project proposals help you to solidify the key selling points of your effort and gain organizational support for your improvement ideas. It also forms a foundation for the development of your project plan, which again, should be developed to better ensure success for your implemented project.

Project Stakeholders

Project stakeholders are included in the project team; however, special attention should be given to them to be sure that you have included your project stakeholder in the project management and implementation process. Some project stakeholders are readily identified. If those project stakeholders that can kill or refuse to use the resulting project are not included in the project management and decision-making project, those individuals can thwart all of the effort that you have put into the project development effort. Take care to identify all of the project stakeholders.

A personal example of this occurred in a consulting study with a creative gift company. The team had assembled those who they thought were the key decision makers in support of the project. As the consulting project began, a key decision maker had not been brought into the previous scoping of the effort. In the first meeting with the corporate decision makers, this individual refused to participate and said that they would not use the system once implemented. The approach and scope of the project needed to be adjusted to provide value to the organization and implement the system. The resulting system, however, did not encompass all of the intended scope.

Another example of this is a project with an airline caterer. A significant amount of effort has been expended to develop an analytical approach to reduce costs for the organization. Senior executives had been engaged in the development effort and the team had developed a system that minimized operational costs. Unfortunately, the CEO of the organization had not been brought into the effort and his objectives were not to minimize costs but were to maximize sales so that the company could go public. The system was revised to meet the CEO's needs, however, this constituted rework from the original effort.

Model Automation

After the metrics, the structure, the data and statistics, and the methodology have been selected, they can then be integrated into a quantitative decision-model. The decision model should be as automated as reasonably possible so that it can be easily utilized on an ongoing basis. The degree to which a model is automated should be based on the end users and their understanding of the decision process. The number of users and the access to the model dictate where it should reside, for instance, on a single machine, a company network, or the Internet. The following are some suggestions for computer tools that may be useful for some decision situations.

- **Basic spreadsheet**—Use a basic spreadsheet for statistical models and data models. Statistical and data analysis capabilities exist within a spreadsheet, and most people easily understand them. Use macros to program basic screen functions and generate analyses to assist the user. Although the spreadsheet is easier to use, it does not have the computational capabilities that more sophisticated models require.
- **Databases**—Use databases with simple queries and macros to automate basic decision models. Again, the level of sophistication of the model is minimal, however; it may be sufficient for solving the problems at hand. This type of model would focus primarily on the use of queries or Standard Query Language (SQL) statements to generate the various summary data and metrics.
- **Databases or spreadsheets with front-end screens and presentation capabilities** — Build user interfaces into the model using database languages for spreadsheets on a limited basis. They provide an additional level of ease to access the data. You can use programming languages to build computer screens that you can design for specific user needs. Also, leverage presentation systems such as PowerBI to provide user friendly results to the decision-makers.

- **Complete computer application systems with data input, data output, and processing logic**—A complex quantitative decision model that has a number of capabilities requires the development of a customized computer application to support the model. Data input screens, data output screens, processing logic, and database support are required as necessary. Also, develop reports to provide the decision maker with the type of analysis and output required to make decisions in a timely manner. These applications can differ in complexity based on end user requirements and the sophistication of the user.
- **Computer application systems on a company network**—Depending upon the number of users, you can develop a network-based decision model that is accessible through a company network. Develop the model so that a number of individuals can simultaneously use the system. More rigorous design, development, and testing would be required for a system that is networked to a larger audience.
- **Computer applications are available from the Internet**—You can also use computer applications and websites to access decision models. A simple example of this may be an Internet site that displays the latest car models for a car manufacturer without query capabilities. This could be compared to an Internet site that provides users with the ability to input criteria for an automobile such as make, year, color, and gas mileage and then present all the cars available and their prices that satisfy these criteria. This approach is applicable for individuals in remote locations.

Automate a quantitative decision-making model only to the degree that it is beneficial in its overall use. For example, a small model used infrequently to make decisions on a local level should not be automated in the same way as decision applications accessible to a national user base. The development of a decision model must be cost effective and consider the time and resources available [3, 4].

Business Cases Development

In various decision processes and analyses, typically there is more than one alternative or potential solution to a problem. Decision makers should fully understand the ramifications of the costs associated with a dollar investment in an alternative. To do this, one must assess the alternative over the time horizon associated with the project life. This involves assessing alternatives in terms of the time value of money, effects of taxation, depreciation, and inflation. This includes assessing alternatives based on their equivalent worth, rate of return, payback period, cost–benefit analysis, and break-even point.

With project initiation, you need to develop a clear business case to support the financial requirements of your project. This involves understanding the outgoing costs and inflowing revenues that are projected for your project. Many organizations have standardized business case development templates. Typically, you work with your finance organization to enable you to develop and complete the business case. You should, however, understand how the project costs and revenues are assessed and the impact of interest rates and time on the development of these business cases.

You must assess the cost of activities within an organization to understand the economic impact of processes, operations, and investments. Costs may be fixed, variable, reoccurring, or have one-time investment costs. Additionally, you must also include costs associated with ongoing maintenance, operations, repair, and disposal when analyzing problem areas. From a holistic perspective, include life cycle costs from inception to disposal. Project management and life cycle costing methodologies work to capture the entire set of related costs associated with an activity. These costs extend to marketing, training, disposal, overhead, and others, which are not normally accounted for in the evaluation process.

You can use a Cost Breakdown Structure to ensure that you capture all the associated costs with a project or activity. This hierarchical diagram identifies typical costs found in a variety of different types of improvement activities or projects. There are four major categories of costs: Research and Development, Investment, Operations and Maintenance, and System Phase-Out and Disposal [5] Capture project costs by category and year to account for each type of cost. Figure 1 shows a corporate level diagram of a Cost Breakdown Structure at a summary level.

You can incur costs at a point in time and at a specified interest rate. You can also generate revenues and cost savings at a point in time. Use these cash flows or series of inflows and outflows of cash over time to measure the profitability of a given activity. Use

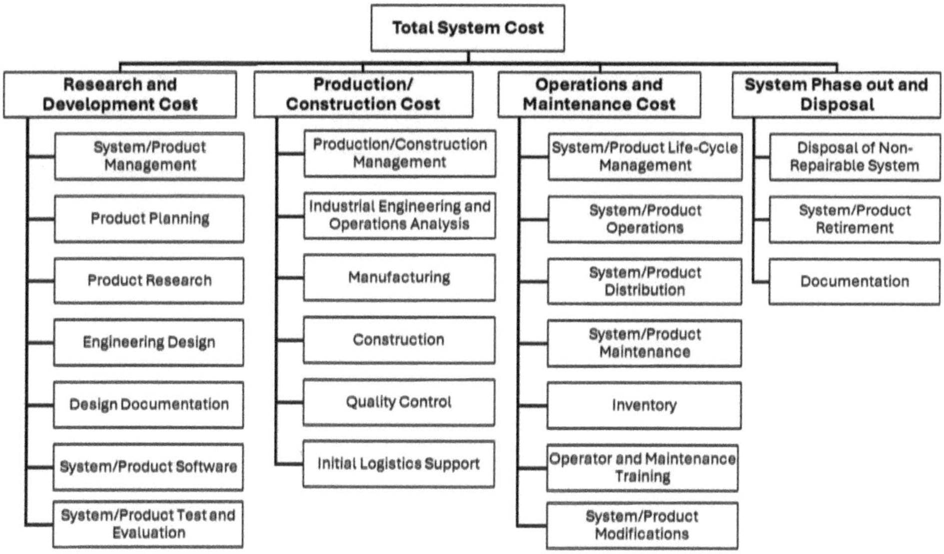

Fig. 1 Cost breakdown structure (adapted from Fabrycky and Blanchard [6])

the time value of money and economic equivalence to measure the inflows and outflows of cash and provide a point of comparison for these measures. Typically, you use the Net Present Value, Rate of Return, or Payback Period to measure the profitability of the project. Numerous books and texts address these important concepts for the analyses of the use of corporate resources. You should account for the total system cost and all of these economic considerations when determining various courses of action.

References

1. Kerzner, Harold, Project Management: A Systems Approach to Planning, Scheduling, and Controlling, 13th Edition, John Wiley and Sons, 2022
2. Verzuh, Eric, The Fast Forward MBA in Project Management, Sixth Edition, John Wiley and Sons, 2021
3. Cassone, Deandra T., Tillman, Frank A., *A Professional's Guide to Problem Solving with Decision Science: An Integrated Approach for Assessing Issues, Finding Solutions, and Reaching Corporate Objectives*, Second Edition Overland Park, KS, Pioneering Partnerships, LLC, January 2018
4. Tillman, Frank A., Cassone, Deandra T., *A Professional's Guide to Decision Science and Problem Solving: An Integrated Approach for Assessing Issues, Finding Solutions, and Reaching Corporate Objectives*, Upper Saddle River, NJ, Financial Times Press/Pearson Education, Web: December 1, 2012, Hardcover: February 2012.
5. Fabrycky, W. and Blanchard, B. Life-Cycle Costs and Economic Analysis, Prentice Hall, Hoboken, NJ, 1991
6. Wolter J. Fabrycky, Benjamin S. Blanchard, Life-Cycle Cost and Economic Analysis Prentice Hall, 1991

Monitoring Success in Accomplishing Objectives

Abstract

Project management techniques to execute the improvement projects are discussed. Different types of project management techniques can be used depending on the nature of the project. Successful project implementation is important to support the quantitative decision-making process. Assuming successful project implementation, monitoring the real impact of the effort should occur. Additionally, using the models and methods that have been implemented, the decision maker should perform sensitivity analysis with the model so that they can test and see areas of further impact in their operations. This involves assessing whether the key operational metrics established earlier are being met and if the organization is being strategically moved in the direction of improvement.

Project Execution

To realize the objectives of the organization and the objectives of the improvement effort, the project to implement the solution must be well executed. This requires the appropriate planning and engagement of your project management team to ensure that the project is delivered on time, on budget and with the needed scope to accomplish it. Anyone that has implemented projects simply knows that this is a challenge. Very few projects are delivered with all these elements intact, however, it is worth spending the time to do the appropriate planning and structured execution to deliver projects well [1].

D. Cassone, *Operational Leadership Using Quantitative Decision Making*,
Synthesis Lectures on Engineering, Science, and Technology,
https://doi.org/10.1007/978-3-031-67565-2_9

Project Planning Based on Project Management Approach

The nature of the project being delivered should be best matched to the appropriate project management approach. These approaches fall into three primary categories which include the waterfall, agile and a hybrid approach. Entire courses and fields of study exist in project management. The primary means to select an approach to project management lies in the degree of uncertainty and ability to define the work required to complete the project. Definable work typically has clear procedures, similar projects, and proven past successes (car, airplane, appliances, building projects, etc.). Exploratory work has high uncertainty, exploratory, high rate of change, new designs (product development, software engineers, doctors, lawyers, etc.) This section will only touch on the approaches briefly as they pertain to the delivery of the identified project [2].

Waterfall Approach

The Waterfall approach is best used with definable work. Fundamental project management tools and techniques including a project charter, stakeholder analysis, statement of work, work breakdown structure and a detailed breakdown of work to be performed at the initiation of the project are employed with the waterfall approach to project management. With the breadth of types of non-software and physical development projects, understanding the fundamentals of waterfall project management is critical.

Agile Approach

Agile is an umbrella term that covers a variety of frameworks and methods. This approach requires an initial high-level definition of the project where requirements are prioritized into a product backlog. Work is performed in iterations where each iteration is structured, the customer selects the most important functionality, the team delivers a working project, daily team meeting are used to coordinate and prioritize actions and, in the end, the customer reviews the product and accepts or iterates on the product functionality. This is repeated for the prioritized backlog of work.

Hybrid Approach

With the hybrid approach, more than one project management approach is used for the project. This involves a combination of traditional and iterative approaches. This combination of approaches can be designed into the project management process and used for overall planning, feature development, etc.

As mentioned above, select the project management approach that best suits the project needs. Combinations of approaches can also be used to support the specifics of the development effort.

The Tripple Constraint—Scope, Schedule, Cost

Scope, schedule, and cost are known as the triple constraint of project management [3]. Key in the delivery of projects are the maintenance of the project schedule, keeping the project within budget, maintaining the key features identified at the outset of the project all at a high quality. Appropriate project management techniques, whether waterfall, agile or hybrid should be used to manage these three elements and keep the project team and stakeholders abreast of progress. No one likes surprises at the end of a project development effort, especially when any of these three key elements are missing or have been exceeded. Good communication and engagement with stakeholders is required to keep the project team informed of progress. As a note and not discussed in any detail in this text, there are robust project management tools, no matter the approach, that should be used to keep your team engaged and productive and keep your stakeholders informed.

All projects contain risks, and these project risks should be identified and planned for in the project management process. The basic concept of project risk management focuses on the proactive management of the known unknowns. Typically, little or no risk management planning is performed. Risks are seen as a bad thing and issues and escalation occur to blame others on the issues. Instead of planning, identifying, analyzing, response planning and controlling risks, they simply aren't identified and result in reactive crisis management for a project. This requires significantly more resources and takes longer for problems to surface.

A robust project management plan should be created to manage the project execution. Key elements of that plan include the project scope, work breakdown structure, schedule baseline, costs baseline, change management plan and a management plan for procurement, schedule, cost, scope, quality, risk, and others. Project closure should also be addressed in the project management plan.

There are numerous project management resources available to manage project execution and the intent in this book is not to deeply discuss project management. Many organizations have their own project management structures and may have a Project Management Office. If you are not in an organization that does, Eric Verzuh, The Fast Forward MBA in Project Management [4] provides downloadable templates and is helpful from a real-world perspective for project planning and execution.

What-If and Sensitivity Analysis

The first answer generated from a quantitative decision-making model is usually the starting point of the decision-making process. A well-defined model has the functional capability to change key parameters and constraints and determine the impact of those changes on the final solution. What-if analysis is a key ingredient in the decision process. In many ways, the information gathered, analyses performed, and models developed all lead to understanding the impact of the decisions made in the changing operating environment. Not only should what-if analysis be performed with the completed model, but it also should be a primary use of the model.

For example, you can easily change budget levels in a resources allocation model because available resources can change at any time. Additionally, the decision maker may want to test funding scenarios with different budgets to determine what projects are the best to be funded now with limited resources. You normally build this what-if analysis into a decision model.

Use sensitivity analysis to determine the degree to which various variables and parameters in the model can change and their impact on the solution. This helps determine the sensitivity of solutions to the variability of the data. Having the ability to perform sensitivity analysis is important to determine the range of parameters that can cause changes in model outcomes. This gives you insight into the accuracy of the data and the variability of parameters and establishes confidence in the results. Both what-if analysis and sensitivity analysis are extremely important capabilities and are the main reasons to build decision models [5, 6].

Monitoring Success

The information gathered from the approach discussed in this book and integrated into a quantitative decision-making model should support the decision-making process. Using this analysis process provides decision makers with data and information to improve the decision-making process whether or not a model is developed [4, 5]. Keep focused on the problem to be solved. The quantitative decision-making model should provide the structure to support and institutionalize a robust decision-making process.

Go back to your original analysis in establishing the goals, objectives, decision criteria and metrics. Re-evaluate the metrics, data and statistical analysis, integration of operational methods and use of decision-making techniques after the approach has been implemented. It will most likely take time to see the results but monitor the strategic direction and operational performance of your organization to determine its impact. A few questions that may be beneficial to ask are listed below.

- Has your organization funded projects in alignment with their goals?
- Are you seeing improvements in your operational performance?
- Are you seeing additional market share with your products?
- Are you spending less money and improving performance overall?
- Do your benchmarks reflect this in your industry?

The goal of the book is to provide you with a quantitative decision-making approach to enhance your organizational performance. Track your original metrics and other industry standard metrics and benchmarks to see if you have been able to move company operations in the right strategic direction.

References

1. PMI Pulse of the Profession® 2023 Report Shows the Value of Power Skills in Project Success, Updated 29 NOVEMBER 2022, Link: PMI Pulse of the Profession®2023 Report
2. Project Management Institute, Inc., Agile Practice Guide, Newtown Square, Pennsylvania, 2017
3. Kerzner, Harold, Project Management: A Systems Approach to Planning, Scheduling, and Controlling, 13th Edition, John Wiley and Sons, 2022
4. Verzuh, Eric, The Fast Forward MBA in Project Management, Sixth Edition, John Wiley and Sons, 2021
5. Adapted from Cassone, Deandra T., Tillman, Frank A., *A Professional's Guide to Problem Solving with Decision Science: An Integrated Approach for Assessing Issues, Finding Solutions, and Reaching Corporate Objectives*, Second Edition Overland Park, KS, Pioneering Partnerships, LLC, January 2018
6. Adapted from Tillman, Frank A., Cassone, Deandra T., *A Professional's Guide to Decision Science and Problem Solving: An Integrated Approach for Assessing Issues, Finding Solutions, and Reaching Corporate Objectives*, Upper Saddle River, NJ, Financial Times Press/Pearson Education, Web: December 1, 2012, Hardcover: February 2012

MIX
Papier aus verantwortungsvollen Quellen
Paper from responsible sources
FSC® C105338

If you have any concerns about our products,
you can contact us on
ProductSafety@springernature.com

In case Publisher is established outside the EU,
the EU authorized representative is:
Springer Nature Customer Service Center GmbH
Europaplatz 3, 69115 Heidelberg, Germany

Printed by Libri Plureos GmbH
in Hamburg, Germany